普通高等院校电子信息类系列教材

通信工程专业"卓越工程师教育培养计划"系列教材

Android 移动开发技术与应用

李学华　王亚飞　编著

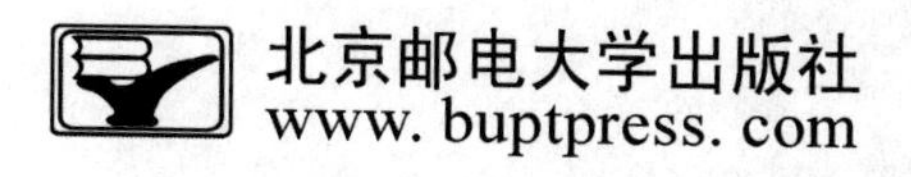

内容简介

本书以生动具体的案例介绍Android移动开发技术，力求通过实际的应用案例使读者快速掌握Android移动开发技术。

全书共分为9章，其中第1～5章介绍Android移动开发的基础；第6～9章介绍4个具体的案例(包括两个应用类和两个游戏类)，这些案例涵盖了Android移动开发中所需要的基本技术和技巧。

本书可以作为高等学校电子信息类专业本科生的教材，也可以作为移动应用开发技术人员的参考书。

图书在版编目(CIP)数据

Android移动开发技术与应用/李学华，王亚飞编著. --北京:北京邮电大学出版社，2013.8

ISBN 978-7-5635-3616-0

Ⅰ. ①A… Ⅱ. ①李…②王… Ⅲ. ①移动终端－应用程序－程序设计 Ⅳ. ①TN929.53

中国版本图书馆CIP数据核字(2013)第179369号

书　　名：Android移动开发技术与应用
著作责任者：李学华　王亚飞　编著
责任编辑：刘　颖
出版发行：北京邮电大学出版社
社　　址：北京市海淀区西土城路10号(邮编：100876)
发 行 部：电话：010-62282185　传真：010-62283578
E-mail：publish@bupt.edu.cn
经　　销：各地新华书店
印　　刷：北京联兴华印刷厂
开　　本：787 mm×1 092 mm　1/16
印　　张：15
字　　数：387千字
印　　数：1—3 000册
版　　次：2013年8月第1版　2013年8月第1次印刷

ISBN 978-7-5635-3616-0　　定　价：30.00元

《Android 移动开发技术与应用》编审委员会

移动通信和互联网成为当今世界发展最快、市场潜力最大、前景最诱人的两大业务，它们的增长速度是任何预测家未曾预料到的。移动互联网是移动通信和互联网二者不断发展、融合的产物。移动互联网的优势决定其用户数量庞大，截至2012年9月月底，全球移动互联网用户已达15亿。

Android是一种基于Linux的自由及开放源代码的操作系统，主要用于移动设备，如智能手机和平板电脑，由谷歌公司和开放手机联盟领导及开发。2011年第一季度，Android在全球的市场份额首次超过塞班系统，跃居全球第一。2013年5月数据显示，Android占据全球智能手机操作系统市场75%的份额，中国市场占有率为90%。

在移动互联网的快速发展以及Android高市场占有率的背景下，基于Android操作系统的移动开发人才供不应求。为适应通信工程“卓越计划”人才的培养要求，打造卓越的移动开发人才，编者从工程实际的角度出发，以案例式教学为指导，特编写本书。本书系统介绍了基于Android操作系统的移动开发技术，并精选了4个实际的应用案例来说明具体开发过程。本书第1～5章介绍了Android移动开发的基础；第6～9章介绍了4个实际的应用案例(包括两个应用类和两个游戏类)，这些案例涵盖了一般开发中所涉及的技术基础和基本技巧，通过对案例的学习能够快速掌握Android移动开发技术的精要，提高学习效率，达到事半功倍的效果。

本书由北京信息科技大学信息与通信工程学院的教师和悦成移动互联网孵化基地的技术人员共同编写。其中第6，7，9章的案例为北京信息科技大学近几年参加北京市计算机应用大赛的一等奖获奖作品，这些作品的作者包括王補平、彭宇文、杨婉秋、王鑫龙、赵业、何昊、晏冉、彭文欢、任宏志、刘亚杰、邹莉娟、杨飞、张春雷等同学，在此向他们表示感谢。

由于时间有限，错误难免，敬请读者指正！

编　者

目录

第1章 3G移动互联网的发展

移动互联网是指互联网的技术、平台、商业模式和应用与移动通信技术结合所产生的业务及活动的总称。本章在分析第三代移动通信(3rd Generation,3G)主流技术的基础上,介绍基于移动互联网的增值业务,对比主流移动终端平台及相关技术介绍各类应用商场及其特点,并重点分析 App Store 模式。

1.1 3G 发展概述

1908 年 5 月 9 日,美国肯塔基州一个叫内森・斯塔布菲尔德的瓜农获得了移动电话专利,发明了地球上第一部移动电话。虽然这部移动电话比他种的瓜还大,但它的出现不仅仅实现了我们随时随地保持沟通的梦想,而且逐步深入了我们的生活,未来将彻底改变我们的工作、生活、学习、娱乐方式。

从内森・斯塔布菲尔德(如图 1-1 所示)的专利到真正可用的移动电话出现,用了 60 多年时间。1973 年摩托罗拉总设计师马丁・库伯(如图 1-2 所示)带领他的团队用 6 周时间完成了世界通信史上的巨大突破,研制出“便携式”移动电话——一个采用数以千计的零件制造而成,仅仅为了实现无线通话功能的机器。随后,他和他团队还制造出了天线,建造了手机基站。这些基站相当于一台微型电脑,可以测量电话信号的强度,同时把较弱的信号传递至下一个通信蜂窝。

图 1-1 移动电话专利持有者内森・斯塔布菲尔德

图 1-2 摩托罗拉总设计师马丁・库伯

手机的诞生只是普及道路上的一小步，从研发成功到推出市场，摩托罗拉公司等了整整10年的时间，注意是“等”，不是用。10年里，摩托罗拉除了建立第一批的手机基站外，就是无奈地等待美国当局用漫长的时间去审批办理这个他们从未见过的怪东西。

从19世纪80年代起，手机开始逐渐流行，手机自问世至今，经历了第一代模拟制式手机（First Generation，1G）和第二代GSM、CDMA等数字手机（Second Generation，2G），而当前通信运营商和终端产品制造商倡导的3G，都是指将无线通信与互联网等多媒体通信结合的新一代移动通信系统。

到目前为止，手机对于每一个移动用户来讲，不仅仅是一个通话工具，而是一部能够满足我们沟通交流、信息获取、娱乐、商务办公等需求的个人移动智能数字终端。

1.1.1 从1G到3G

1. 1G

1G是指第一代移动通信系统，其代表为模拟移动网。

在20世纪80年代初期，个人通信系统（Personal Communication System，PCS）开始流行，当时有许多国家提倡应该要在个人通信上实现移动化，为了实现这个需求，各种不同的通信技术纷纷被提出，例如早期的卫星电话。当时无线基站建置成本相当高昂，卫星电话等许多空中无线通信平台都是被看好的技术，但是最后普及的却是美国的AMPS系统。AMPS系统最早是在1970年代美国开始试验，后来在1981年斯堪的纳维亚开始了商业化服务，而日本在1980年也开始了商业运转。

1983年6月13日，摩托罗拉终于推出名为Dyna TAC 8000X的手机——第一台商用的移动终端，它重794 g，长33 cm，标价3 995美元，最长通话时间是一个小时，可以储存30个电话号码。笨重厚实的深刻印象使美国人称之为“鞋机”，而国人（中国人）习惯称之为“大哥大”，因为它真的很大，甚至可以用来防身。

在美国，随着AMPS系统不断地改良，到1990年，AMPS系统已经成为了一个全国性的服务。其最大特色是采用了细胞式（或称蜂窝式，因为每一个基站彼此间服务的范围紧邻而成一个细胞网络）系统，可以允许每个人在一个基站的服务范围内拨出和接收电话。这是现今所有移动电话网络的始祖，即使到现在全世界的移动电话还是采用地面蜂窝网络的架构，丝毫没有改变。AMPS系统即是我们所俗称的1G。一直到1990年，AMPS系统趋于饱和，于是世界各地许多公司提出其他的解决方案。北美提出两种解决方案，欧洲提出一种。北美两种方案分别是NA-TDMA与CDMA系统，欧洲提出的是GSM系统。这3种系统即是俗称的2G。

2. 2G

2G是指第二代移动通信系统，其代表为GSM。2G以数字语音传输技术为核心。GSM全名为移动通信全球系统“Global System for Mobile Communication”。1982年欧洲的AMPS系统已经进入了商业部署，但是欧洲当局预期移动电话将会有长足的进步，于是促使“欧洲邮电管理会议”研究新的移动电话技术。该会议成立了一个特别移动电话小组（Group Special Mobile or Special Mobile Group），而该小组的成员以开头的第一个字母自称GSM，这就是GSM的由来。GSM主要设计的目的是可以国际漫游，这也是其他标准如 CDMA IS-95 没有的优点。而 GSM 也不同于 AMPS 系统，它是纯数字的信号，而 AMPS 是模拟信号，所以 GSM 可提供较高的数据传输服务带宽，最高可达9 600 bit/s，在当时已经是很够用了。GSM 因为目标放在国际漫游上，所以后来全球市场将近 70%都是 GSM 系统，而采用 CDMA 标准的国家如美国、加拿大、韩国这些国家，因为无法与其他国家互通而被有些用户冠上了“通信孤岛”之类的称号，这是 GSM 非常成功的一点。NA-TDMA 是美国所提出的另一个 AMPS 改进方案。

2.5G 是基于 2G 与 3G 之间的过渡类型。比 2G 在速度、带宽上有所提高。可使现有 GSM 网络轻易地实现与高速数据分组的简便接入。

目前商业应用的 2.5G(Generation)移动通信技术是从 2G 迈向 3G 的衔接性技术，突破了 2G 电路交换技术对数据传输速率的制约，引入了分组交换技术，从而使数据传输速率有了质的突破，是一种介于 2G 与 3G 之间的过渡技术。2.5G 的出现主要是由于 3G 是个相当浩大的工程，所牵扯的层面较多且复杂，要从目前的 2G 一下迈向 3G 是不可能马上实现的。2.5G 移动通信技术的代表有 GPRS、HSCSD、WAP、EDGE、蓝牙(Bluetooth)、EPOC 等。

2G 手机除了可以通话，还可以进行数据业务，例如短信、手机报、手机上网等。

3. 3G

3G 是指第三代移动通信系统。3G 是将无线通信与互联网等多媒体通信结合的新一代移动通信系统。它能够方便、快捷地处理图像、音乐、视频流等多种媒体形式，提供包括网页浏览、电话会议、电子商务等多种信息服务。为手机融入多媒体元素提供强大的支持。但为了提供这种服务，无线网络必须能够支持不同的数据传输速度，也就是说在任何环境中能够分别支持至少 2 Mbit/s、384 kbit/s 以及 144 kbit/s 的传输速度。2G 网络提供的带宽是 9.6 kbit/s。2.5G 增加到 56 kbit/s。3G 将具有更宽的带宽，其传输速度将达到 100～300 kbit/s，不仅能传输话音，还能传输数据，从而提供快捷、方便的无线应用。

在经过了 AMPS 与 GSM、CDMA(Code Division Multiple Access)与 NA-TDMA 等系统的改进后，国际电信联盟(International Telecommunication Union，ITU)在 1990 年就提出 3G 的概念，称为 IMT-2000 标准(International Mobile Telecommunication 2000)。但是这个标准只是提出一个大略的描述，说明 3G 应该具备什么样的特性，达到什么样的要求。到了 1998 年，ITU 接受 15 个关于 IMT-2000 的技术标准建议案，其中采用卫星的有 6 个，9 个采用地面基站建议案(又是一次卫星与地面技术之争)。而地面建议案中，有 8 种采用 CDMA 相关技术，所以 CDMA 几乎是成为 IMT-2000 标准的主流。

而后的两个重要组织确定了这些标准之中何者出线。一个是 3GPP，另一个是 3GPP2，各组织成员纷纷在 ITU 提案希望可以争取关于标准之制订。一直到了 1999 年 11 月，从所有相关建议案中选出了 4 项技术，分别是 WCDMA(Wideband Code Division Multiple Access)、cdma2000(Code Division Multiple Access 2000)、UTRA TDD、EDGE。EDGE 是 IS-136 的升级版本，UTRA TDD 后来变成了 TD-SCDMA(Time Division-Synchronous Code Division Multiple Access)。3G 的标准演变就此确定。

1998 年 12 月，由欧洲电信标准协会 ETSI 发起，日本 ARIB、TCC，美国 T1 与韩国 TTA 参与的第一个 3G 跨国团体成立，称为 The Third Generation Partnership Project，

简称3GPP。这就是3GPP的由来。3GPP在空中无线电技术上采用两种技术，速度高而广域的环境中使用WCDMA，小区域而低速环境下使用TD-CDMA(URTA TDD)，核心网络采用GSM系统。后来中国的大唐集团将其TD-SCDMA申请成为3GPP的标准方案，TD-CDMA的主要支持者西门子于是与大唐合作将TD-SCDMA合力申请成为正式的选择方案。目前3GPP主要的两个选择方案确定为WCDMA与TD-SCDMA。

与3GPP相对应的是3GPP2，由美国国家标准协会(ANSI)，日本ARIBO、TTC，韩国TTA与中国企业发起，主要厂商就是Qualcomm等CDMA相关厂商为主，支持cdma2000，核心网络采用CDMA IS-95。

1.1.2 3G主流技术标准分析

国际电信联盟(ITU)早在2000年5月即确定了WCDMA、cdma2000和TD-SCDMA3个主流3G标准。下面介绍一下3G相关的主要技术。

1. cdma2000

cdma2000也称为CDMA Multi-Carrier，由美国高通北美公司为主导提出，摩托罗拉、朗讯和韩国三星都已参与，韩国现在成为该标准的主导者。这套标准是从窄频cdma2000 1X数字标准衍生出来的，可以从原有的cdma2000 1X结构直接升级到cdma2000 3X(3G)，建设成本低廉。但目前使用CDMA的地区只有日本、韩国和北美，中国联通也应用了该模式过渡。不过cdma2000的技术研发却是目前各标准中进度最快的，许多3G手机也已率先面世。

CDMA技术一向是高通公司独大的局面，与cdma2000一样，目前有能力做出cdma2000核心组件的只有高通公司，而且所有的重要专利也掌握在其手中。目前，cdma2000 1X已经在韩国正式大规模商业运作。

在中国，中国电信的3G网络采用cdma2000制式。

2. WCDMA

WCDMA是采用宽频分码多重存取技术，是由GSM网发展出来的3G技术规范，这套系统能够架设在现有的GSM网络上，对于系统提供商而言可以较方便地过渡，而GSM系统相当普及的亚洲对这套新技术的接受度会比较高。因此，WCDMA具有先天的市场优势。

除了编码是采用CDMA技术外，WCDMA系统的核心网络(如交换机等)还是采用GSM系统，只有手机与基站需要更新。所以WCDMA不可以视为是CDMA的升级，反倒是应视为GSM的升级，因为WCDMA的支持者几乎都是GSM的相关业者。

为了可以同时在GSM基站与WCDMA基站收到信号，出现了双模手机(Dual-mode Terminals)的概念：即在2G(GSM系统)与3G(WCDMA)之间提供非手动无感觉的模式切换、越区切换及漫游。

WCDMA是目前电信运营商采用最多且最为成熟的3G制式。在中国，联通的3G网络采用WCDMA制式。

3. TD-SCDMA

该标准是由我国大唐电信公司提出的3G标准。该标准将智能无线(Smart Antenna)、同步 CDMA 和软件无线电等当今国际领先技术融于其中。由于中国国内庞大的市场，该标准受到各大主要电信设备厂商的重视，全球一半以上的设备厂商都宣布可以支持 TD-SCDMA 标准，对于中国通信事业实为一大机遇。TD-SCDMA 是一个混合多个标准的技术，可视为是 TD-CDMA、SCDMA、SDMA 的综合体。大唐集团是 TD-SCDMA 主要的研发单位，在设备开发方面，大唐的方针是在自主开发核心技术的基础上，广泛对外合作，走联合开发之路。

在标准与基地站系统开发方面，大唐 1999 年开始与西门子合作。在手机终端方面，大唐将和 TI、诺基亚、LG、普天集团等十家国内外企业共同成立一个股份公司，联合进行 TD-SCDMA 手机的芯片、IC 及手机的参考设计。大唐电信科技集团于 2001 年 9 月 20 日与中电东方通信研究中心有限公司(CECW)、飞利浦半导体 3 家公司连手签署了 TD-SCDMA 终端方面合作意向书，并宣布将以成立合资公司的方式共同开发 TD-SCDMA 终端芯片。大唐将向新公司提供 TD-SCDMA 技术解决方案；飞利浦将提供 3G 核心芯片开发所用的标准通信平台；CECW 则将提供终端有关的协议软件和终端测试技术。

在中国，中国移动的 3G 网络采用 TD-SCDMA 制式。

1.2 蓬勃发展的增值业务

增值电信业务是相对于基本电信业务而言的。电话、电报、用户电报、传真和数据传输等使用公众电信网，直接为用户提供信息传送的业务，称为基本电信业务。

增值电信业务是利用基本电信网的资源，配置计算机硬件、软件和其他一些技术设施，并投入必要的劳务，使信息的收集、加工、处理和信息的传输、交换结合起来，从而向用户提供基本电信业务以外的各式各样的信息服务。由于这些业务是附加在基本电信网上进行的，起增加新服务功能和提高使用价值的作用，因而称做增值电信业务。

简而言之，移动增值业务就是指我们每个手机用户除了打电话之外的，需要支付费用的业务，统称为增值业务，报刊短信、彩信、彩铃、手机上网、移动阅读、手机支付等。

1.2.1 增值业务发展概述

1991 年芬兰开通了全球第一个商用的 GSM 全球移动通信系统，意味着数据业务可以应用于移动网络，1994 年短信服务出现于芬兰，拉开了移动增值业务的大幕，1998 年芬兰开通 Jippii 平台，提供铃音下载，从此移动增值业务开始蓬勃发展。

目前我国移动通信运营商开放的主要增值业务如下。

(1) 短信

短信(SMS)是指用户可以在移动电话上直接发送和接收的文字或数字消息。用户一次能接收和发送短消息的字符分别为 140 个英文或数字字符，或 70 个中文字符。

(2) 彩信

彩信是移动运营商推出的多媒体消息服务(MMS)，能够支持多媒体功能，传递功能全面的内容和信息，这些信息包括文字、图像、音频、视频、数据等各种多媒体格式的信息。彩信与原有的普通短信比较，除了基本的文字信息以外，更配有丰富的彩色图片、声音、动画、振动等多媒体内容。

(3) 彩铃

彩铃是指由被叫用户定制,用户设定一款回铃音后,当主叫用户拨打被叫用户手机时为主叫用户提供一段悦耳的音乐或一句问候语来替代普通回铃音。

(4) 手机上网

手机上网具有"实时在线"、"按量计费"、"快捷登录"、"高速传输"、"自如切换"等特点,随时满足用户上网聊天、移动炒股、在线游戏的无限需求,给用户的工作生活带来更多实惠和便捷。

(5) 移动音乐

移动音乐是移动运营商推出的一种通过内置于手机终端的音乐用户端软件实现的业务。它是一种新型的音乐体验消费平台,用户通过移动音乐,可以在线收听运营商提供的全曲音乐,也可以订购彩铃、下载振铃。

(6) 手机阅读

手机阅读是移动运营商通过多样化的阅读形式向用户提供各类电子书内容,以在线和下载为主要阅读方式,让用户尽享无线阅读新体验。

(7) 手机支付

手机支付业务是移动运营商面向用户提供的综合性移动支付服务,用户开通手机支付业务,系统将为用户开设一个手机支付账户,用户可通过该账户进行远程购物(如互联网购物、缴话费、水费、电费、燃气费及有线电视费等)。

(8) 移动应用商场

移动应用商场(Mobile Market)是聚合各类开发者及其优秀应用和数字内容,满足多类型终端用户实时体验、下载、订购需求的综合商场,通过手机用户端、WWW 网站和 WAP 网站为用户提供软件、游戏、主题、视频、音乐、图书等一站式服务。

另外还有手机邮箱、生活资讯、手机游戏、手机电视等。

1.2.2 从增值业务到数据业务

增值业务最初的含义是附加在电信主营业务之上增加价值的部分,但是随着增值业务在运营商收入中所占比重的逐步增大,运营商慢慢发现,在未来,增值业务将是运营商收入的主要部分,再叫"增值业务"就不太合适了,确切的说法是"数据业务"。

数据业务提升运营商的利润水平,给用户带来了更多的方便。在未来,用户会选择更多数据业务来方便自己的工作与生活。用户获取数据业务的方式也在发生变化,主要体现之一便是"应用商场"模式。

1.3 移动终端技术

1.3.1 智能手机介绍

功能手机(Feature Phone)是指那些不能随意安装、卸载软件的普通手机,一般只具有手机自带的通信及相关功能。

传统手机使用的是生产厂商自行开发的封闭式操作系统，所能实现的功能非常有限，不具备智能手机的扩展性。自从Java出现以后，使“功能手机(Feature Phone)”逐渐具备了安装Java应用程序的功能，但是当时这种扩展了的功能手机的用户界面操作友好性、运行效率及对系统资源处理，都远远不及“智能手机(Smart Phone)”。

智能手机比传统的手机具有更多的综合处理功能。智能手机同传统手机外观和操作方式类似，不仅包含触摸屏，也包含非触摸屏数字键盘手机和全尺寸键盘操作的手机。“智能手机(Smart Phone)”就是一台可以随意安装和卸载应用软件的手机(就像电脑那样)。3G时代下，智能手机已成主流，智能手机市场发展迅猛。IDC日前发布的数据显示，2010年，制造商们共出货智能手机3.05亿台，2010年第四季度，全球智能手机出货量超越PC，成为里程碑式标志；2011年智能手机出货量达4.72亿台，增长率达55%；到2015年全球智能手机出货将达9.82亿台。正如IDC高级分析师Kevin Restivo所说，“智能手机的闸门已经打开”，智能手机成了一种大趋势。

智能终端除了包含智能手机外，还包含平板电脑。平板电脑界的明星产品为iPad，目前已推出两代。Android平板电脑增长迅速，另外HP也推出了基于RIM系统的平板电脑、Intel的MeeGo平台也瞄准了平板电脑市场。

2009年1月7日我国3G牌照发放，这一事件标志着我国的3G移动互联网产业正式进入大发展阶段。尽管3G解决了网速过慢的问题，但3G移动互联网要想有大的发展，同样离不开智能手机、智能手机操作系统的发展，也离不开应用软件的发展，2011年智能手机应用爆发，成为中国的移动互联网元年。

智能手机操作系统的竞争格局不断变化，市场研究公司IDC的调查数据显示，截至2013年5月，Android以75%的市场份额遥遥领先，苹果公司的iOS以17.3%位居第二，微软公司的Windows Phone发展稳健，以7%位居第三，而前两年风光无限的BlackBerry和Symbian则一落千丈，分别跌至2.9%和0.6%的低谷。

智能手机的功能特点如下：

(1) 具有开放性的操作系统，可以安装更多的应用程序，使智能手机的功能可以得到无限扩展。

(2) 具备无线接入互联网的能力，各种2G、3G网络制式以及WiFi。

(3) 具有PDA的功能，包括PIM(个人信息管理)，日程记事，任务安排，多媒体应用，浏览网页。

(4) 人性化，可以根据个人需要扩展机器功能。

(5) 功能强大，可扩展性能强，可支持的第三方软件多。

智能手机的配置特点包括：

(1) 高速度处理芯片。智能终端一般需要处理音频、视频，甚至要支持多任务处理，这需要一颗功能强大、低功耗、具有多媒体处理能力的芯片。

(2) 大存储芯片和存储扩展能力。

(3) 面积大、标准化、可触摸的显示屏。

(4) 支持播放式的手机电视。以现在的技术，如果手机电视完全采用电信网的点播模式，网络很难承受，而且为了保证网络质量，运营商一般对于点播视频的流量都有所控制，因此，广播式的手机电视是手机娱乐的一个重要组成部分。

(5) 支持 GPS 导航。它不但可以帮助你很容易找到你想找到的地方，而且 GPS 导航还可以帮助找到你周围的兴趣点，未来的很多服务也会和位置结合起来，这是智能手机特有特点。

(6) 操作系统必须支持新应用的安装。有可能安装各种新的应用，使用户的手机可以安装和定制自己的应用。

(7) 配备大容量电池，并支持电池更换。3G 无论采用何种低功耗的技术，电量的消耗都是一个大问题，必须要配备高容量的电池，1 500 mA·h 是标准配备，随着 3G 的流行，未来外接移动电源也会成为一个标准配置。

(8) 良好的人机交互界面。

1.3.2 Symbian OS 介绍

1. Symbian OS 平台概述

1998 年 6 月，Psion 公司联合手机业界巨头诺基亚、爱立信、摩托罗拉等组建了 Symbian 公司。该公司继承了 Psion 公司 EPOC 操作系统软件的授权，并且致力于为移动信息设备提供一个安全可靠的操作系统和一个完整的软件及通信器平台。

作为一种开放式平台，任何人都可以为支持 Symbian 的设备开发软件。这意味着开发伙伴具有更多可供选择的应用，同时拥有更大的市场。为此 Symbian 推出了白金合作计划吸引了大量的厂商加入。Symbian 公司还大量参与 WAP、Wireless Java 和 Bluetooth 的制定工作，确保 EPOC 将完全支持市场的内容和服务需求模块化、可伸缩性、低能耗以及与 Strong ARM 这类 RISC 芯片的兼容性。诺基亚全资收购 Symbian 公司并宣布开源计划，将 Symbian 操作系统开源，使得 Symbian 成为一个开放的、可扩展的智能手机平台。

Symbian OS 系统按照人机交互界面大致分为 S60、S80、UIQ 等。不同的用户界面对应不同的手机和模拟器屏幕尺寸、分辨率以及不同的输入方式。其中：S60、S80 等对应的手机是采用键盘输入方式；UIQ 对应的手机采用触摸屏方式与用户交互。2008 年诺基亚推出的 S60 5th 手机和诺基亚 5800 XpressMusic 加入了对触摸屏的支持。

2010 年由 Symbian 基金会开发的 Symbian 3 已经在之前的 Symbian 平台之上进行了升级，整合了 Symbian OS 各种界面，推出的手机包括 N8、C7 和 C6-01 等机型。

2. Symbian OS 开发环境

开发 Symbian 平台的手机软件，可以采用多种开发工具，如微软研发的 Visual C++ 6.0/Visual Studio 2005、飞思卡尔(Freescale)半导体公司推出支持多种硬件平台的集成开发环境 CodeWarrior，或者是诺基亚研发的 ADT(Application Developer Toolkit)集成开发环境工具包，ADT 的目标是为开发手机应用软件的开发者提供方便的开发环境，其中集成了 Carbide. C++，可以用来开发 Symbian S60 应用程序。

需要安装的软件如下：

(1) Java SDK;

(2) Active Perl(使用5.6.1系列版本,其他版本可能产生不兼容现象,导致搭建环境失败);

(3) Application Developer Toolkit (ADT)(包含 Carbide. C++IDE);

(4) Symbian S60 Platform SDK(包含:编译工具、模拟器及开发帮助文档)。

依次安装完后,即可启动 ADT 中的 Carbide. C++集成开发环境,进行 Symbian 项目开发。

另外,诺基亚扩展了 Qt 开发库,推出了 Nokia Qt SDK,其中也包含了集成开发环境以及 Symbian 平台应用软件开发的 SDK 等软件,可以用来开发 Symbian 平台的应用程序。

1.3.3 Android平台介绍

1. Android 平台概述

谷歌于 2007 年 11 月宣布,与 30 多家业内企业成立开放手机联盟(Open Handset Alliance,OHA),共同开发 Android 开源移动平台。Android 也是一款智能手机操作系统,Android 其实是一个操作系统的称谓,它是谷歌在 2005 年收购的一家手机软件公司名,并用 Android 来命名这个全新的操作系统。Android 向手机厂商和手机运营商提供了一个开放的平台,供它们开发创新性的应用软件。Android 是基于 Linux 技术,由操作系统、用户界面和应用程序组成,允许开发人员查看源代码,是一套具有开放源代码性质的手机终端解决方案。

谷歌的 Android 平台现在宣布公布源代码,并允许所有手机厂商加入开发,免费使用,这无疑让手机企业和第三方软件企业都为之振奋。谷歌宣称 Android 联盟成员目前有 34 家,其中芯片制造商包括英特尔、高通、德州仪器、Nvidia 公司;手机制造商包括摩托罗拉、三星、LG 和宏达(HTC);运营商包括中国移动,美国的 Sprint 和 T-Mobile,美国、日本的 NTT DoCoMo 和 KDDI,10 个欧洲国家的 T-Mobile 等,再加上做应用层面的谷歌、SkyPop。截至 2011 年 6 月,Android 集合了 36 家 OEM 厂商,215 家移动运营商,和超过 45 万名开发者。目前,总计有 20 万个应用在 Android 市场里。

2008 年 10 月谷歌的 G1 手机正式推出。该手机是第一款采用谷歌 Android 操作系统的手机。由于 Android 的开放性吸引了众多手机制造商,HTC、摩托罗拉、三星、LG、华为、联想、酷派等手机制造商不断推出 Android 新手机,截至 2012 年 9 月,Android 设备超过 1 亿台,新增日激活量 130 万台。

2. Android 开发环境

Android 开发采用的集成开发环境是 Eclipse,需要具备的工具如下:

(1) JDK 1.6+;

(2) Android SDK 1.6;

(3) Android SDK Setup;

(4) Eclipse IDE for Java Developers。

1.3.4 Windows Phone 平台介绍

1. Windows Phone 平台概述

Windows Phone 系列操作系统是在微软计算机的 Windows 操作系统上变化而来的，因此，早期 Windows Mobile 的操作界面与 Windows 的操作界面非常相似。Windows Mobile 系列操作系统具有功能更强大，多数具备了音频、视频文件播放、网上冲浪、MSN 聊天、电子邮件收发等功能。而且，支持该操作系统的智能手机多数都采用了英特尔嵌入式处理器，主频比较高，另外，采用该操作系统的智能手机在其他硬件配置（如内存、储存卡容量等）上也较采用其他操作系统的智能手机要高出许多，因此性能比较强劲，操作起来速度会比较快。

但是，此系列手机也有一定的缺点，如因配置高、功能多而产生耗电量大、电池续航时间短、硬件采用成本高等缺点。Windows Mobile 系列操作系统包括 SmartPhone 以及 Pocket PC Phone 两种平台。Pocket PC Phone 主要用于掌上电脑型的智能手机，而 SmartPhone则主要为单手智能手机提供操作系统。

Windows Phone 7 是微软推出的一个触控操作模式操作系统，其新特性总结如下：

（1）触摸手势。与 iPhone 类似，滑过、移动、拖曳等。

（2）运动手势。有些是 iPhone 所没有的功能，不会可以使用一系列回旋和加速，相反，它可以使用手机内置相继探测移动并建立适宜的动作。例如，摇晃、扭曲以及电话和物体的相对移动等；当电话放置在一个平面时，可以执行一些动作，并且会判断其自身是否处于口袋中。

（3）具备令人激动的锁屏功能。可以把玩、拖曳、摇动以及旋转等。

（4）具有全新的外观。界面更像是 Windows Vista 的黑色并具有未来主义的视觉效果，支持图形过渡、精细的效果和其他更为华丽的 UI。

（5）主要为手指操作设计。让手机更适合手指操作，单手就可以很轻松地使用并易于理解；尽量取消按钮，使用触摸屏即可执行多数任务。

（6）具备更好的视频回放。媒体播放器和照片管理器得到很大改进，浏览器以全屏运行、具备标签浏览。

（7）键盘得到改进。计划设置全触摸键盘。

2. Windows Phone 开发环境

直接到微软的网站可以下载开发环境所需要的软件安装包。注意，如果开发 Windows Mobile 7 的应用程序，需要在 Windows 7 中进行，安装 Visual Studio 2010 Express for Windows Phone CTP 即可，其中包含了以下组件：

（1）Visual Studio 2010 Express for Windows Phone CTP；

（2）Windows Phone Emulator CTP；

（3）Silverlight for Windows Phone CTP；

（4）XNA Game Studio 4.0 CTP。

1.3.5 iOS 平台介绍

1. iOS 平台概述

iOS 是苹果公司为 iPhone 开发的操作系统，它主要是给 iPhone、iPod Touch 以及

iPad 使用。就像其基于的 Mac OS X 操作系统一样,它也是以 Darwin 为基础的。原本这个系统名为 iPhone OS,直到 2010 年 6 月 7 日 WWDC 大会才宣布将其改名为 iOS。iOS 的系统架构分为 4 个层次:核心操作系统层(the Core OS layer)、核心服务层(the Core Services layer)、媒体层(the Media layer)、Cocoa 界面服务层(the Cocoa Touch layer)。系统操作占用大概 240MB 的存储器空间。

iOS 的用户界面的概念基础是能够使用多点触控直接操作。控制方法包括滑动、轻触开关及按键。与系统交互包括滑动(swiping)、轻按(tapping)、挤压(pinching)及旋转(reverse pinching)。此外,通过其内置的加速器,可以令其旋转设备改变其 *y* 轴以令屏幕改变方向,这样的设计令 iPhone 更便于使用。屏幕的下方有一个 home 按键,底部则是 dock,有 4 个用户最经常使用的程序的图标被固定在 dock 上。屏幕上方有一个状态栏能显示一些有关数据,如时间、电池电量和信号强度等。

2. iOS 开发环境

Cocoa Touch 是从 Mac OS X 上的 framework 裁剪和修改而来,用于开发 iPhone、iPod、iPad 上的软件。也是苹果公司针对 iPhone 应用程序快速开发提供的一个类库。此库以一系列框架库的形式存在,支持开发人员使用用户界面元素构建图像化的事件驱动的应用程序。iPhone 上的 Cocoa Touch 与 Mac OS X 上的 Cocoa 和 AppKit 类似,并且支持在 iPhone 上创建丰富、可重用的界面。

苹果公司为 iOS 开发人员准备了 iPhone SDK,当然 iPhone SDK 只能基于苹果的 MAC OS 系统进行开发。iPhone SDK(Software Development Kit,软件开发包)包括了界面开发工具、集成开发工具、框架工具、编译器、分析工具、开发样本和一个模拟器。

(1) Xcode

Xcode 是 iPhone 开发工具库中最为重要的一款工具。它提供了一个全面的项目开发和管理环境,包括源文件编辑、丰富的文档和一个图形化调试器。Xcode 由多款开源 GNU 工具构建而成,即 GCC(编译器)和 GDB(调试器)。

(2) Interface Builder

Interface Builder (IB)提供了一个快速的原型工具,可用于以图形化的方式布局用户界面以及从 Xcode 源代码链接到这些预构建的界面。借助 IB,可以使用可视设计工具绘制界面,然后将这些屏幕元素连接到应用程序中的对象和方法调用。

(3) 模拟器(Simulator)

iPhone 模拟器运行于 Macintosh 之上,借助它,无需连接到实际的 iPhone 或 iPod touch,就可以在台式机上创建和测试应用程序,当然,并不是 iPhone 所有的特性模拟器都可以模拟。模拟器提供的 API 与在 iPhone 上使用的 API 相同,并针对概念设计的效果提供相应的预览。在使用模拟器时,Xcode 将编译在 Macintosh 上运行的 Intel x86 代码,而不是 iPhone 上使用的基于 ARM 的代码。

(4) Instruments

Instruments 用于分析 iPhone 应用程序的内部运行原理。它对内存利用率进行采样,并监视性能。这样,你可以准确识别并锁定应用程序中的问题区域,并采取有效措施。Instruments 提供基于时间的图形化性能图(plot),可显示应用程序中占用资源最多的地方。Instruments 由 Sun Microsystems 开发的开源 DTrace 包构建而成。Instruments 在

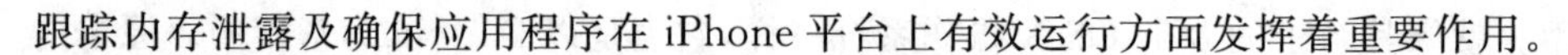

跟踪内存泄露及确保应用程序在 iPhone 平台上有效运行方面发挥着重要作用。

1.3.6 其他平台介绍

其他移动平台还有很多，包括诺基亚和英特尔宣布推出的免费移动平台操作系统 MeeGo，将用于智能手机与平板电脑；Palm 公司（被惠普收购）推出的 Web OS（又称 Palm OS），吸引了无数人的眼球；RIM 公司研发的黑莓手机操作系统 BlackBerry OS；三星公司自行研发的智能手机平台 Bada（于 2009 年 11 月 10 日发布），支持丰富功能和用户体验的软件应用，特点是配置灵活、用户交互性好、面向服务，非常重视 SNS 集成和地理位置服务应用。

1.4 APP Store 模式介绍

App Store（Application Store），通常理解为应用商场。App Store 最初是苹果公司专门针对 iPhone 创建的软件应用商场，向 iPhone 的用户提供第三方的应用软件服务，这是苹果开创的一个让网络与手机相融合的新型经营模式。App Store 现在为 iPhone、iPod Touch、iPad 终端提供应用程序下载。用户通过自行绑定的信用卡来进行付费，付费之后的应用，将自动安装到用户的设备上。

应用商场中的商品并非苹果提供，苹果只是开放针对 iOS 的应用开发包（SDK），以便第三方应用开发人员开发针对 iPhone、iPodTouch 及 iPad 的应用软件。第三方应用开发人员申请成为苹果的开发者之后，可以开发基于 iOS 的应用，开发完毕上传到 APP Store，开发者可以自行为其开发的应用定价，通过 APP Store 审核之后，开发者享有产品收入 70%的分成。APP Store 的完整产业链如图 1-3 所示。

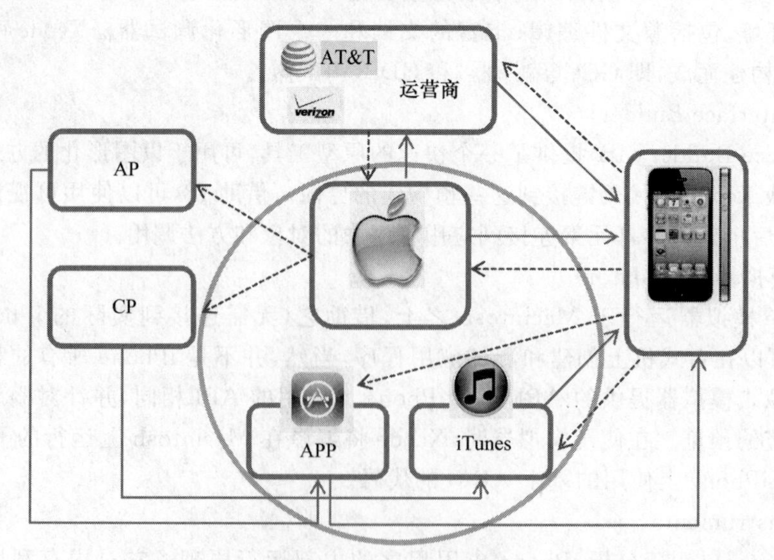

图 1-3 APP Store 产业链

图 1-3 为 APP Store 的产业链，苹果提供 APP Store 和 iTunes 平台，同时在 iPhone、iPad、iPod Touch 中内置 APP Store 客户端，用户点击 iOS 设备中的 APP Store 客户端，

便可看到各类产品的更新，用户选择需要的产品后，会直接从用户信用卡中扣除相应的费用，并通过运营商网络直接将用户选择的应用下载安装至用户设备。苹果公司会将受益者与开发者(AP 或 CP)按 3∶7 比例分成。

App Store 模式的意义在于：为第三方软件的提供者提供了方便而又高效的一个软件销售平台，使得第三方软件的提供者参与其中的积极性空前高涨，适应了手机用户们对个性化软件的需求，从而使得手机软件业开始进入了一个高速、良性发展的轨道，是苹果公司把 App Store 这样的一个商业行为升华到了一个让人效仿的经营模式，苹果公司的 App Store 开创了手机软件业发展的新篇章，App Store 无疑将会成为手机软件业发展史上的一个重要的里程碑，其意义已远远超越了"iPhone 的软件应用商场"的本身。

1.5 移动应用商场分类与分析

应用商场模式顺应了移动互联网时代的趋势，目前移动应用商场非常多，手机厂商、移动运营商、移动平台开发商以及其他机构与企业也推出了第三方应用商场。

1.5.1 手机厂商类应用商场

在 2009 年年底，手机应用商场的概念迅速风靡起来，各大手机厂商开始搭建自己的应用商场，来提升自身手机产品的卖点和吸引力。手机应用商场里的内容涵盖了手机软件、手机游戏、手机图片、手机主题、手机铃声、手机视频等几类。目前国内名气较大的应用商场有以下几家。

1. 苹果软件应用商场(App Store)

2008 年 3 月 6 日，苹果对外发布了针对 iPhone 的应用开发包(SDK)，供免费下载，以便第三方应用开发人员开发针对 iPhone 及 Touch 的应用软件。根据苹果公司 2011 年 6 月公布的数据，目前 iTunes 商场每年的运营费用为 13 亿美元。音乐下载次数为 150 亿次，图书下载次数为 1.3 亿次，应用下载次数为 140 亿次；向开发者支付 25 亿美元分成费用；信用卡账号总数为 2.25 亿个；应用总数 42.5 万个，其中 iPad 应用为 9 万个，游戏和娱乐应用为 10 万个。

2. 三星手机应用商场(Samsung Apps)

三星应用商场是专为使用三星手机的用户打造的一个支持所有操作系统的开放式平台，用户可登录商场选取中意的软件，目前可提供多款免费的游戏、娱乐软件。三星手机用户既可以通过 PC 下载三星应用商场的用户端，并将用户端程序移至手机内存卡中进行安装，也可通过手机的 GSM 网络、3G 网络或 Wi-Fi 直接将用户端下载到手机中进行安装，从而登录三星应用商场。三星出厂的每款智能手机都会预装 Samsung Apps 客户端。

3. 诺基亚软件应用商场(Ovi Store)

在移动通信大会上，诺基亚正式发布了其在线软件和媒体商场"Ovi Store"，该业务将会与 2010 年 5 月在 9 个国家正式上线。"Ovi Store"将提供应用程序、游戏、视频、Widget 小工具、播客(视频分享)、基于地理位置的应用等各种内容，用户可以通过 S60 和

S40 平台手机登录该商场。“Ovi Store”商业模式为，软件开发者将自己开发的软件和应用发布，供其他用户下载，诺基亚将会把来自“Ovi Store”收入的 70%分给软件开发者。

4. 黑莓软件应用商场(BlackBerry App World)

2009 年 4 月 1 日，RIM 应用程序商场正式亮相，取名为黑莓应用程序世界(BlackBerry App World)。黑莓应用程序世界里包含了游戏、办公、娱乐、新闻、天气、保健、社交网络等各种应用。

同苹果 iPhone 应用程序商场运营模式一样，RIM 也将同外部开发者共享收入，即 RIM 应用程序商场每售出一款产品，外部开发者将获得其中 80%的收入，但苹果相应比率为 70%。

5. LG 软件应用商场(LG Application Store)

2009 年 7 月 14 日，LG 宣布正式推出测试版的手机应用程序商场，LG 应用商场将提供支持 15 种语言的 1 400 项应用，其中 100 个程序为免费。它是一项针对 LG 高端手机和智能手机用户的增值服务。同时，与其他手机在线商场不同的是，LG 并不对上线的应用程序种类进行审查，而只是保证它们一流的质量和兼容性。

6. 其他手机厂商应用商场

目前基本所有手机厂商都推出了应用商场，例如联想的“应用商店”、华为的“智慧云”等。

1.5.2 移动运营商类应用商场

各运营商也推出了各自的应用商场。

1. 中国移动软件应用商场

2009 年 8 月 17 日，中国移动应用商场(Mobile Market，MM)正式开门迎客，鼓励开发者开发游戏、软件、主题等多种手机应用商品。移动应用商场是聚合各类手机应用开发者及其优秀应用，满足所有类型的手机用户实时体验、下载和订购需求的综合商场。

Mobile Market 是中国移动在 3G 时代搭建的增值业务平台，由中国移动数据部负责运营，并由广东移动和卓望科技负责共同建设。Mobile Market 平台的运作流程，是用户通过用户端接入运营商的网络门店下载应用，开发者通过开发者社区进行应用托管，运营商通过货架管理和用户个性化信息进行分类和销售。

Mobile Market 是由中国移动投资建设，通过与国内外数百名知名尖端手机软件 CP 合作，面向数亿的移动用户，致力于打造手机终端软件市场百亿级产业链，满足智能手机用户不断提高的安全、创新等需求，聚集并辅导手机终端软件开发商及个人独立开发者发掘终端软件市场需求，进行快速开发并完成安全签名认证，最终发布产品并实现赢利的手机应用软件下载平台。

在应用商品价格方面，开发者开发的应用商品可在中国移动制订的价格范围内自主定价，中国移动进行商品价格确认。在发展初期，Mobile Market 销售的应用支持按次计费，资费区间为 0～15 元/次。后续，将支持包月、按道具收费等多种计费方式。

此外，中国移动将征集的在线应用分为 4 类：A 类软件主要功能必须通过网络实现；B 类软件为离线功能业务，但会不定期数据更新包；C 类是运行过程中可自动触发终端通

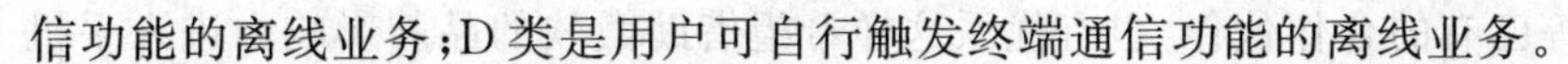

信功能的离线业务；D类是用户可自行触发终端通信功能的离线业务。

2. 中国电信软件应用商场(天翼空间)

2010年3月17日，中国电信天翼空间应用商场正式商用，天翼空间是中国电信为用户提供各类手机应用、数字内容发现、下载、购买的一项移动业务服务。天翼空间由面向用户的应用商城和面向开发者的应用工厂(开发者社区)组成。

其中，应用商城提供的应用软件涵盖影音娱乐、新闻资讯、游戏、理财、实用工具、书籍、旅行、社交网络等类别，包括数百款免费体验、付费购买的产品。天翼应用工厂是为应用开发者提供信息聚集，开发资源，测试资源，销售支持，客服支持以及开发信息交流的服务平台。

天翼空间在为手机用户提供开放、丰富及便捷的手机软件和应用下载的同时，也为手机应用软件提供商和个人开发者提供了完整而成熟的商业应用模式。

天翼空间率先给出了移动互联网发展的新模式：全网、全终端、全开放。这种选择权“还给”用户，给用户最大程度自由选择权的方式，直接促使注册用户量、开发者的数量、应用开发量、下载量和店铺数量呈现出前所未有的快速增长，并带来了品牌形象的大幅提升。截至2011年3月，天翼空间的累计注册用户已经超过了1 000万人，下载量突破1 000万次，软件应用数量达到15 000个。

3. 中国联通软件应用商场(联通沃商店)

2010年11月10日，联通沃商店正式发布，沃商店提供各种手机应用下载，包括手机游戏下载，手机工具下载，手机娱乐下载，手机主题下载，手机生活下载，手机阅读下载等。

联通沃商店与中国移动Mobile Market以及中国电信的天翼空间采用手机扣费方式收费，用户在商店注册时，会自动注册“沃账户”这一中间账户，同时可绑定话费支付或者支付宝、财付通等第三方支付工具，并可以通过一卡通、银行卡两种方式对中间账户进行充值。

1.5.3　移动平台商类应用商场

按照移动平台划分的应用商场有以下几个。

1. 谷歌软件应用商场(Android Market)

谷歌针对苹果的“iPhone App Store”开发了自己的Android手机应用软件下载店“Android Market”，它允许研发人员将应用程序在其上发布，也允许Android用户随意将自己喜欢的程序下载到自己的设备上。根据AndroLib网站的数据，截至2011年5月月底，谷歌Android Market应用商场中免费和付费应用的数量已经达到20万款。

Android Market中的应用数量仍远少于苹果App Store，后者中的应用达到10万款。在移动互联网的使用量方面，iPhone和Android已超过面市较早的智能手机平台。此外，iPhone和Android还创造了新的手机广告经济。

2. 微软软件应用商场(Windows Marketplace)

2009年10月15日，微软在中国市场正式推出最新的手机操作系统Windows Mobile 6.5。全新的手机界面、更人性的用户体验，得到了大批手机厂商的青睐，十余家手机厂商纷纷在第一时间推出首批Windows Mobile 6.5手机。其中既包括多普达、LG、中兴、华为、TCL 等已有的合作伙伴，也有优派这样希望借助微软的品牌影响力进入手机市场的新兴企业。

Windows Mobile 6.5 中内置 Windows Marketplace，向用户提供软件下载服务。

目前 Marketplace 同样支持 Windows Phone 7 手机。

1.5.4 第三方应用商场

常见的第三方应用商场包括如下几个。

1. TOMPDA 手机应用商场

该商场按照移动平台和手机厂商的方式，对智能手机软件进行了分类。其包括 Android、Symbian、Windows Mobile 等多个手机平台软件下载，还提供了按照手机厂商方式的下载分类列表，方便用户下载手机软件。而且网站设置论坛，方便用户对手机软件进行评价讨论，分享使用心得体会。

2. 91 手机应用中心

提供各个主流智能手机平台软件下载，涵盖了 iPhone、Android、Symbian 及微软等手机平台的软件下载。提供手机同步用户端软件"91 手机助手"，方便用户进行手机与 PC 的数据同步。

3. 国内第三方 Android 应用商场

用"雨后春笋"来形容现在国内的 Android 应用商场一点不为过。有一定知名度的安卓应用商场有做论坛出身的老字辈"安卓市场"、"安智市场"、"机锋市场"、"eoeMarket"，有依托 PC 客户端的"应用汇"、"91 手机助手"，还有独立客户端"飞流下载"、"爱米软件商店"等。这些国内第三方 Android 商场的繁荣，给国内 Android 用户获取、安装、使用软件带来了很大的帮助。

第2章 Android概述及平台搭建

本章介绍 Android 平台的系统架构，包括体系结构、系统库、应用程序框架和应用程序，特别是 Android 开发环境的搭建过程。

2.1 Android 概述

2.1.1 系统概述

Android 中文意思为“机器人”，它是美国谷歌公司在 2007 年 11 月 5 日主导推出的一个手机操作系统。该操作系统基于 Linux 内核，且完全开源和免费，到 2011 年年初的数据显示，仅正式发布 4 年的 Android 系统已经超越称霸 10 年的 Symbian（塞班）系统，成为全球最受欢迎的智能手机平台。

Android 由开放手机联盟（Open Handset Alliance）共同研发，该联盟是美国谷歌公司与众多科技公司组建的一个全球性的联盟组织。开放手机联盟包括手机制造商、手机芯片厂商和移动运营商几大类，联盟在成立之初就有 34 位成员，其中包括 HTC、摩托罗拉、三星、LG、中国移动、华为等知名公司。

图 2-1 中列出的机构均为开放手机联盟成员。

图 2-1　开放手机联盟成员

开放手机联盟与谷歌一起来开发 Android 操作系统及其应用软件，共同开发Android 的开源移动系统，它们都在 Android 平台的基础上不断创新，让用户体验到最优质的服务，这使得 Android 具有强大的生命力和竞争力。

2.1.2 系统特性

Android 之所以成为万众瞩目的国际巨星，有其特有的优点：

(1) 开放源代码。Android 最大的特性是源代码全部开放，可以从谷歌的官方网站上免费下载到 Android 系统的所有源代码。这是以前所有手机操作系统中从来没有过的，而开放手机联盟，致力于共同制定标准使 Android 成为一个开放式的系统。

(2) 应用广泛。Android 系统除了可以安装在手机这样的终端设备，还可以把 Android 操作系统安装到像 PAD、车载导航仪 GPS、MP4，包括一些笔记本电脑这些硬件上，应用非常广泛。

(3) 可扩展性强。Android 系统里面内置了谷歌特有的业务，比如搜索、导航、gmail、google talk 语音搜索等，而在 Android 上所有应用都是可替换和可扩展的，即使核心组件也是一样。可以充分发挥想象力，创造出自己的 Android 王国。

(4) 云计算。云计算最早是谷歌倡导并推动的一项新的技术，未来将没有服务器概念，平时所用的电脑都将作为存储数据的云端。Android 设备在未来也会成为云端的一个设备。

(5) 硬件调用。Android 内置了重力感应器，加速度感应器，温度、湿度感应器等硬件传感器，另外 GPS 模块，WiFi 模块，也让更多的硬件调用更加方便。

(6) 开发方便。Eclipse＋ADT＋Android SDK 的开发环境，非常容易集成，开发和调试也更加方便快捷，另外，由于 NDK 的支持，使得 C 和 C＋＋核心算法更容易加入到开发程序中来。

除此之外，Android 在对 Web 的支持上，支持最新的 HTML5 和 JavaScript 脚本；Android 不断更新的 SDK，使得虚拟键盘和多点触碰等成为可能；Android 的个性支持，在 Widget、Shortcut、Live Wallpapers 上体现出华丽和时尚。

Android 的特点还有很多，其未来让人充满期望。

2.1.3 硬件特性

作为一个使用 Linux 内核的智能手机操作系统，Android 的 CPU 至少应为 ARM9 200 MHz，这样才能带动 Dalvik 这个 Java 级虚拟机。谷歌官方最早推出的 G1 手机使用的是 ARM11 和 ARM9 组成的双核 CPU，主频达到了 520MHz。虽然 Linux 内核在内存消耗方面有一定的优势，但是 Android 桌面、UI 等都工作在 JVM 之上，需要占用的内存很大，在 T-Mobile G1 上达到了 192 MB，比使用本地 C/C＋＋编写的程序更占用资源。同时，由于 Android 程序生命周期的特殊性，GC(Garbage Collection)不会频繁地回收资源，所以占用的内存非常大。

在 3D 硬件加速方面，可以由厂商自己来订制，作为一个可选的组件来支持 OpenGL ES，最新已经支持到了 2.0 以上。厂商还可以订制 WiFi 网卡、各种感应器、摄像头等硬件配置,为 Android 系统提供强大的支持。

最新的 Android 3.0 的硬件标准要求是屏幕分辨率和双摄像头,要求屏幕达到 1 280×800 像素,配有前后两个摄像头即可。而双核处理器,将会通过硬件兼容性解决。

2.2 Android 系统架构

2.2.1 体系结构

Android 系统是基于 Linux 和 Java 技术,它在底层采用 Linux 内核和本地库,在上层提供 Java 支持框架和开发接口。它借助于 Linux 强大的稳定性、开放性和可移植性,Java 语言开发的广泛性、简单性和可移植性,一经推出就受到广泛关注和欢迎,在嵌入式开发中产生比较深远的影响。通过上一节的介绍,对 Android 已经有了初步的了解。下面将介绍 Android 的体系结构,如图 2-2 所示。

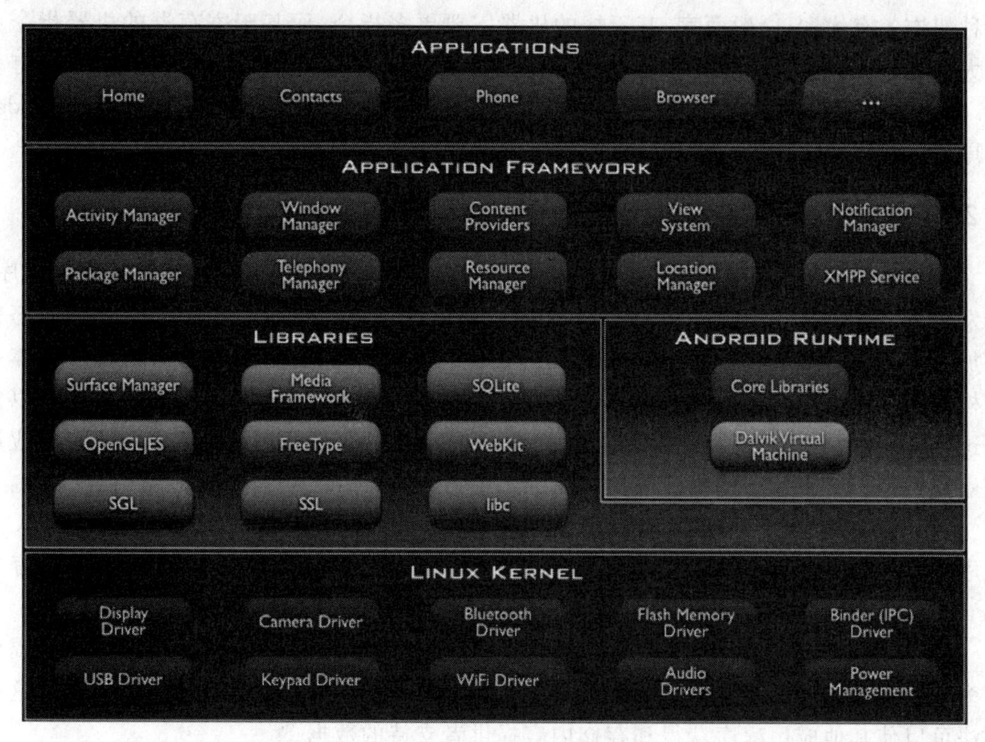

图 2-2 Android 的体系结构图

要了解 Android 的整个体系结构,这张图是非常重要的。由图中可以看出 Android 体系分为 5 个部分,从下至上,依次是 Linux 内核层(Linux Kernel),Android 运行时(Android Runtime)、核心库(Libraries)、应用框架层(Application Framework)和应用层(Applications),下面我们将对这几层进行简单的介绍。

2.2.2 系统库

此层包括核心库与运行时两部分：

(1) Libraries (核心库)

Android 包含一个 C/C++库的集合，供 Android 系统的各个组件使用。这些功能通过 Android 的应用程序框架(application framework)暴露给开发者。这些包括系统 C 库(标准 C 系统库(libc))媒体库、界面管理库、图形库、数据库引擎、字体库等。

(2) Android Runtime (Android 运行时)

Android 包含一个核心库的集合，提供大部分在 Java 编程语言核心类库中可用的功能。每一个 Android 应用程序是 Dalvik 虚拟机中的实例，运行在它们自己的进程中。Dalvik 虚拟机设计成，在一个设备可以高效地运行多个虚拟机。Dalvik 虚拟机可执行文件的格式是.dex，dex 格式是专为 Dalvik 设计的一种压缩格式，适合内存和处理器速度有限的系统。大多数虚拟机包括 JVM 都是基于栈的，而 Dalvik 虚拟机则是基于寄存器的。两种架构各有优劣，一般而言，基于栈的机器需要更多指令，而基于寄存器的机器指令更强大。dx 是一套工具，可以将 Java .class 转换成 .dex 格式。一个 dex 文件通常会有多个.class。由于 dex 有时必须进行最佳化，会使文件大小增加为原来的 1～4 倍，以 odex 结尾。Dalvik 虚拟机依赖于 Linux 内核提供基本功能，如线程和底层内存管理。

2.2.3 应用程序框架

通过提供开放的开发平台，Android 使开发者能够编制极其丰富和新颖的应用程序。开发者可以自由地利用设备硬件优势、访问位置信息、运行后台服务、设置闹钟、向状态栏添加通知等。开发者可以完全使用核心应用程序所使用的框架 APIs。应用程序的体系结构旨在简化组件的重用，任何应用程序都能发布它的功能且任何其他应用程序可以使用这些功能(需要服从框架执行的安全限制)。这一机制允许用户替换组件。所有的应用程序其实是一组服务和系统。应用程序框架包括：

(1) Activity Manager(活动管理器)。Activity 是 Android 的核心类，它相当于 C/S 程序中的窗体或 Web 的页面，而 Activity Manager 用来负责管理当前程序中所有的 Activity从创建直到销毁的全过程。

(2) Content Provider(内容提供者)。为其他应用程序提供数据，也就是说，它提供了多个应用程序之间共享数据。在 Content Provider 中定义了一系列的方法，通过这些方法可以使其他应用程序获得和存储内容提供者支持的数据。

(3) Notification Manager(通知管理器)。使应用在状态栏显示自定义的警报通知，通知可以用很多种方式来吸引用户的注意力——闪动背灯，震动，播放声音等。一般来说是在状态栏上放一个持久的图标，用户可以打开它并获取消息。

(4) Resource Manager(资源管理器)。提供对非编码资源的访问通道，例如本地化字符串、图形、布局文件、音频视频等，对于这些资源，Android 会进行分类编译管理。

(5) Location Manager(定位管理器)。提供对 GPS、基站等信息的获取以提供用户的位置信息，确定当前的手机用户的位置。

(6) Telephony Manager(电话语音模块)。管理电话、语音等设备。

(7) View System(显示框架)。Android 系统各种显示控件管理,由显示框架提供。

2.2.4 应用程序

Android 装配一个核心应用程序集合,包括电子邮件客户端、SMS 程序、日历、地图、浏览器、联系人和其他设置。所有应用程序都是用 Java 编程语言写的。开发者可以直接调用这些应用,也可以利用此模式分享自身的 API。而对于运营商而言,可以借此嵌入自身的增值应用,同时开放其 API,建立自己的软件生态圈。

如果想获取更多的优秀软件,或者,让他人分享自己的程序设计,还有一个更方便的选择:Android Market,这是世界上第一个出售 Android 应用程序的在线商店,它由谷歌创办,地址如下:

```
http://www.android.com/market
```

在 Android 手机上可以通过 Android Market 客户端浏览和下载商店中的应用程序(其中有免费和收费两类程序),如果已经开发好一个 Android 应用程序,只需要在 Android Market 上注册(25 美元),就可以上传自己的作品了。在这里,介绍一下 Android 比较常用的系统库:

(1) SGL。2D 图形引擎。它的主要作用就是做 2D 图形的渲染。比如我们平时玩的一些 2D 游戏,游戏里边会涉及一些图片,包括文字、图形等,这些内容的绘制实际都是由这个引擎所提供的,受这个引擎的支持。

(2) OpenGL ES。OpenGL 为开放式图形库。它提供了性能卓越的 3D 图形标准,该套标准,跨编程语言,跨操作系统。而 OpenGL ES 是专为嵌入和移动设备设计的 3D 轻量图形库,它基于 OpenGL API 设计,该库可以使用硬件 3D 加速或者使用高度优化的 3D 软加速。

(3) Webkit。浏览器引擎。它支持 Android 浏览器和一个可嵌入的 Web 视图。现在 Android 系统里的浏览器用的是它,包括 iPhone 的浏览器也用的是 Webkit,Webkit 是一个非常成熟的浏览器的引擎。

(4) SQLite。数据库引擎。它是一个优秀的轻量级关系型数据库。SQLite 虽然是轻量级,但是在执行某些简单的 SQL 语句时甚至比 MySQL 和 Postgresql 还快,在 Symbian,Mobile,Android 系统中都支持嵌入式数据库,也就是 SQLite,它是 Android 存储方案的核心。

(5) 媒体库。基于 PacketVideo OpenCORE,该库支持多种常用的音频、视频格式回放和录制,同时支持静态图像文件。这些库支持播放和录制许多流行的音频和视频格式,以及静态图像文件,包括 MPEG4、H.264、MP3、AAC、AMR、JPG、PNG。

2.3 Android 开发环境搭建

工欲善其事,必先利其器,下面将详细介绍如何搭建 Android 的开发环境,这是学习 Android 应用开发必备的基础知识之一。

2.3.1 软件准备与安装

开发 Android 应用至少需要具备如下开发工具和开发包：

(1) Java SE SDK(简称 JDK,Java 标准开发工具包)；

(2) Eclipse (集成开发工具)；

(3) Android SDK (Android 标准开发工具包)；

(4) ADT(Android Development Tools 开发 Android 程序的 Eclipse 插件)。

2.3.2 开发环境配置

在安装这些工具和开发包之前,我们先需要下载它们,具体如下所示。

(1) 下载 JDK(http://java.sun.com/javase/downloads/index.jsp)。注意,要选择与自己的系统相匹配的版本,本书中选择 JDK 6,下载界面如图 2-3 所示。

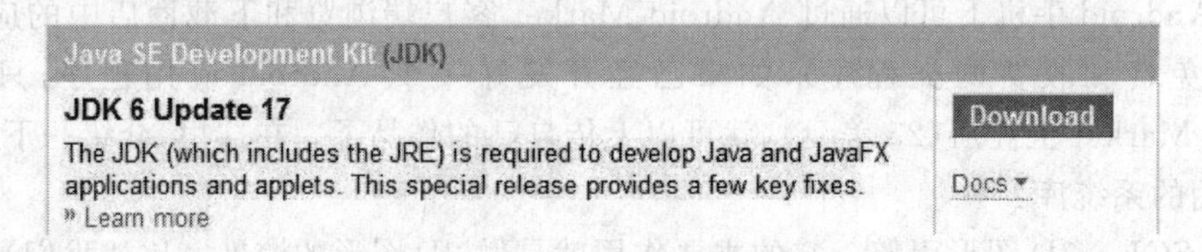

图 2-3 JDK 下载界面

(2) 下载 Eclipse(http://www.eclipse.org/downloads/)。推荐使用 Eclipse Classic 版本或者 Eclipse IDE for Java EE Developers 版本,本书选择"eclipse-Classic 3.6.2-win32.zip",下载界面如图 2-4 所示。

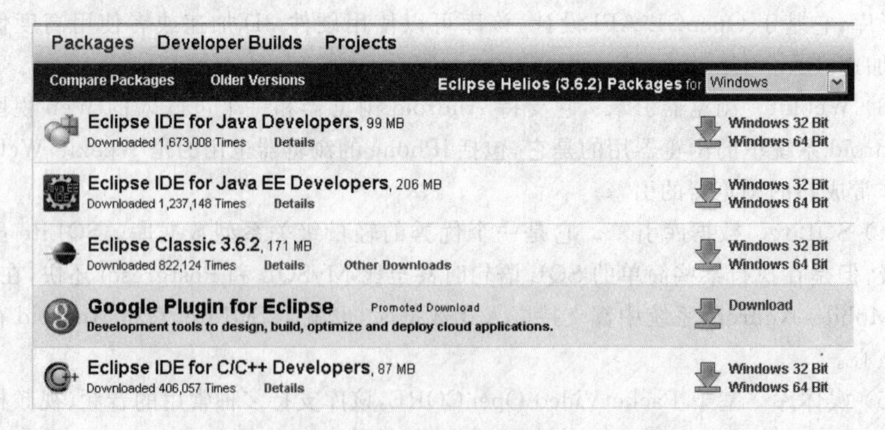

图 2-4 Eclipse 下载界面

(3) 下载最新的 Android SDK(http://developer.android.com/sdk/index.html)。为了安装方便,这里选择"android-sdk_r11-windows.zip",下载界面如图 2-5 所示。

(4) 安装 JDK。直接运行下载后的安装文件,根据提示进行安装即可。

(5) 设置执行路径。在完成了 JDK 的安装以后,还要设置一个附加步骤,把 jdk/bin 目录添加到执行路径中。所谓执行路径是指操作系统搜寻本地执行文件的目录列表。对于不同的操作系统，这个步骤有所不同，我们着重介绍 Windows 下的配置方法。

Platform	Package	Size	MD5 Checksum
Windows	android-sdk_r11-windows.zip	32837554 bytes	0a2c52b8f8d97a4871ce8b3eb38e3072
	installer_r11-windows.exe (Recommended)	32883649 bytes	3dc8a29ae5afed97b40910ef153caa2b
Mac OS X (intel)	android-sdk_r11-mac_x86.zip	28844968 bytes	85bed5ed25aea51f6a447a674d637d1e
Linux (i386)	android-sdk_r11-linux_x86.tgz	26984929 bytes	026c67f82627a3a70efb197ca3360d0a

图 2-5　Android SDK 下载界面

右击“我的电脑”，选择“属性”选项，如图 2-6 所示。

图 2-6　选择“属性”选项

在打开的对话框中选择“高级”选项卡，单击“环境变量”按钮，如图 2-7 所示。

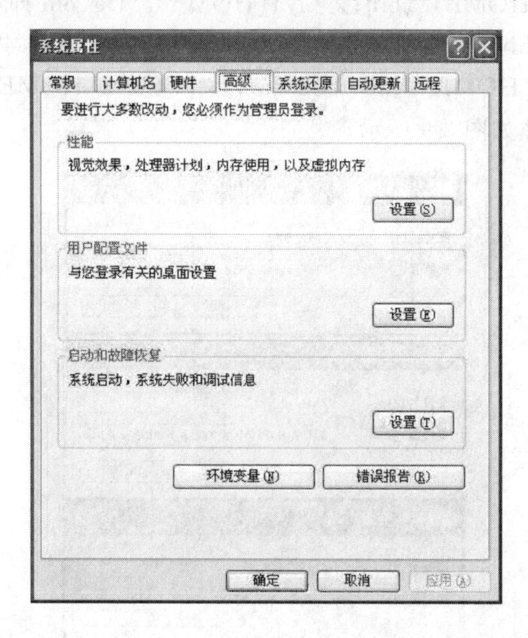

图 2-7　打开环境变量

打开的对话框如图 2-8 所示，在“系统变量”中设置 3 项属性：Java_HOME、PATH、CLASSPATH（大小写无所谓）。若已存在，则单击“编辑”按钮；若不存在，则单击“新建”按钮。

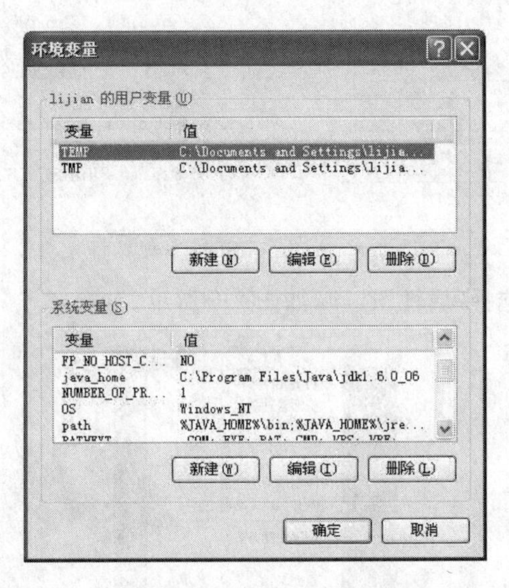

图 2-8　设置环境变量

Java_HOME 指明了 JDK 的安装路径，就是刚才安装时所选择的路径 C:\Program Files\Java\jdk1. 6 0_06，此路径下包括 lib、bin、jre 等文件夹（此变量最好设置，因为以后运行 tomcat 和 eclipse 等都需要依靠此变量）。Path 使得系统可以在任何路径下识别 Java 命令，设为%Java_HOME%\bin;%Java_HOME%\jre\bin 即可。CLASSPATH 为 Java 加载类（class or lib）路径，只有类在 classpath 中，Java 命令才能识别，设为%Java_HOME%\lib;%Java_HOME%\lib\tools. jar，其中%Java_HOME%就是引用前面指定的 Java_HOME。具体如图 2-9 所示。

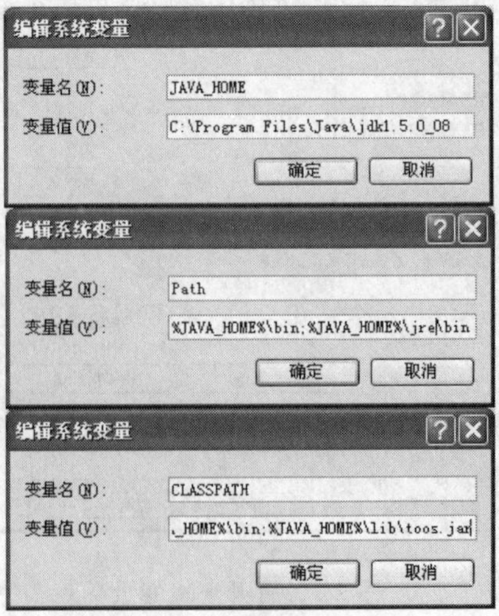

图 2-9　设置系统变量的值

（6）检查配置。

① 依次单击“开始”→“运行”，输入“cmd”，进入命令行模式，如图 2-10 所示。

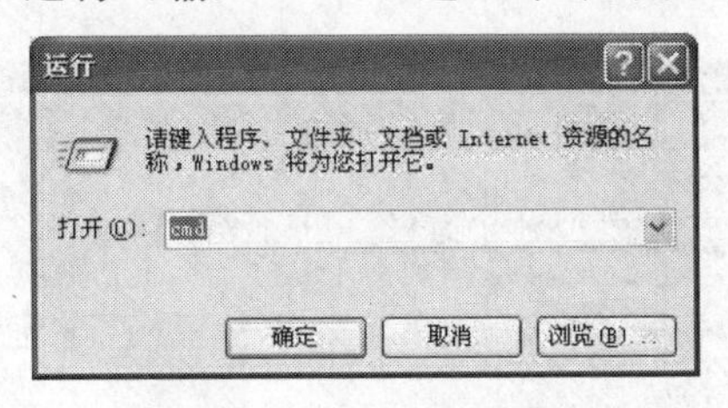

图 2-10　进入命令行模式

② 在弹出的命令行窗口中输入命令“java -version”，如果出现图 2-11 所示的画面，则说明环境变量配置成功。

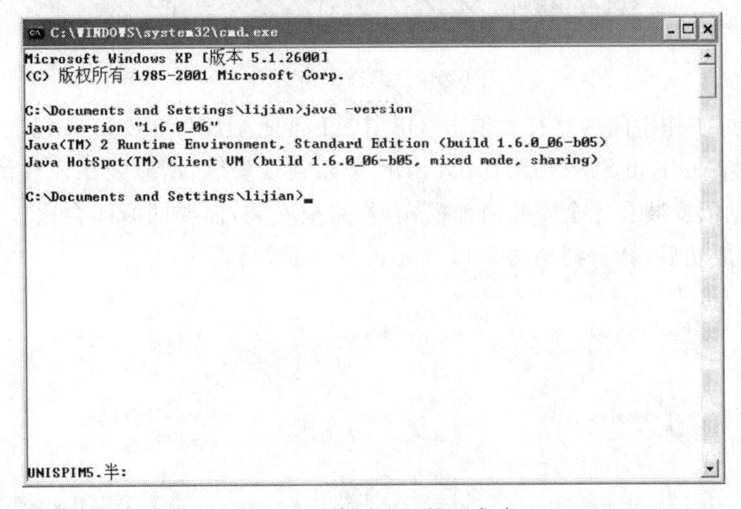

图 2-11　检测配置是否成功

（7）安装 Eclipse。

① 前面我们已经下载了 Eclipse 的安装程序“eclipse-SDK-3.6.2-win32.zip”，无须执行安装程序，此压缩文件解压后即可使用。

② 进入解压后的目录，可以看到一个名为“eclipse.exe”的可执行文件，双击此文件直接运行，Eclipse 能自动找到用户先期安装的 JDK 路径，启动界面如图 2-12 所示。

图 2-12　Eclipse 的启动画面

③ 如果用户是第一次安装和启动 Eclipse，将会看到选择工作空间的提示，如图 2-13 所示。

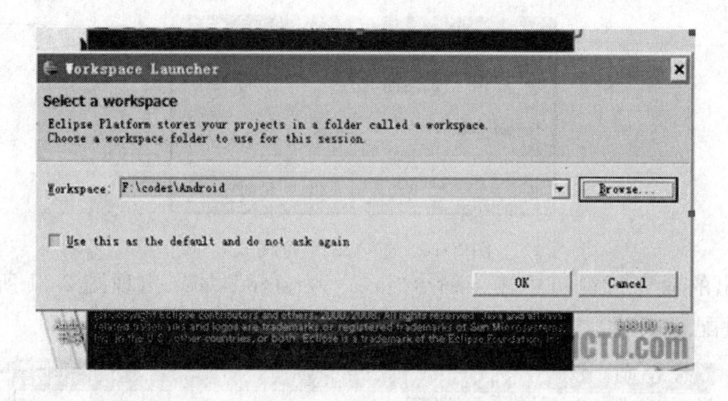

图 2-13　选择工作空间

④ 选择好工作空间的路径后单击“OK”按钮，至此，JDK 和 Eclipse 已经安装完毕。

(8) 安装 Android SDK。Android SDK 必须在线安装，所以要保证有稳定而快速的网络。有自动以安装和完全安装两种模式，若完全安装，需要的时间会比较长，需要耐心等待，等安装成功后，会看到如图 2-14 所示的根目录结构。

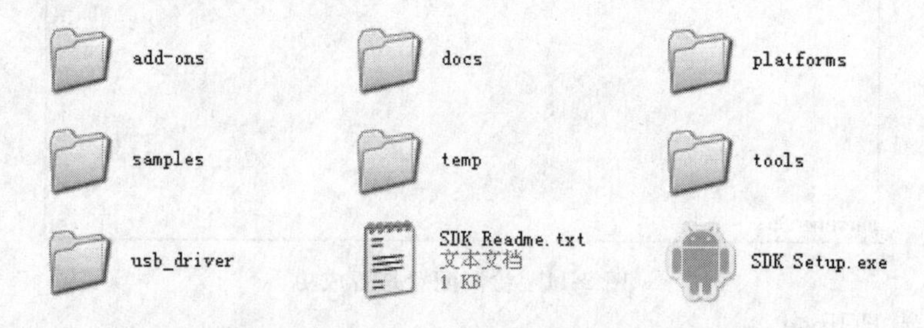

图 2-14　Android SDK 的根目录结构

(9) 安装 Eclipse 插件 ADT。

① 在 Eclipse（版本 3.4）中，依次单击“Help”→“Install New Software...”→“Add...”菜单选项，在弹出的对话框中输入如下地址 https://dl-ssl.google.com/android/eclipse，如图 2-15 所示。

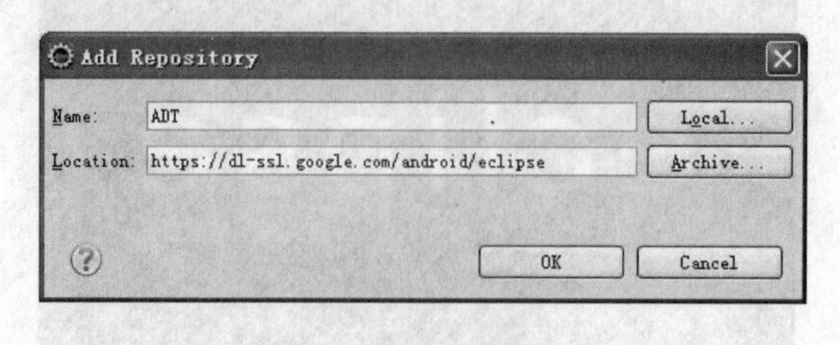

图 2-15　添加 ADT 插件

② 单击“OK”按钮后，回到 Install（安装）页面，选中 Android DDMS 和 Android Development Tools 两项，如图 2-16 所示。

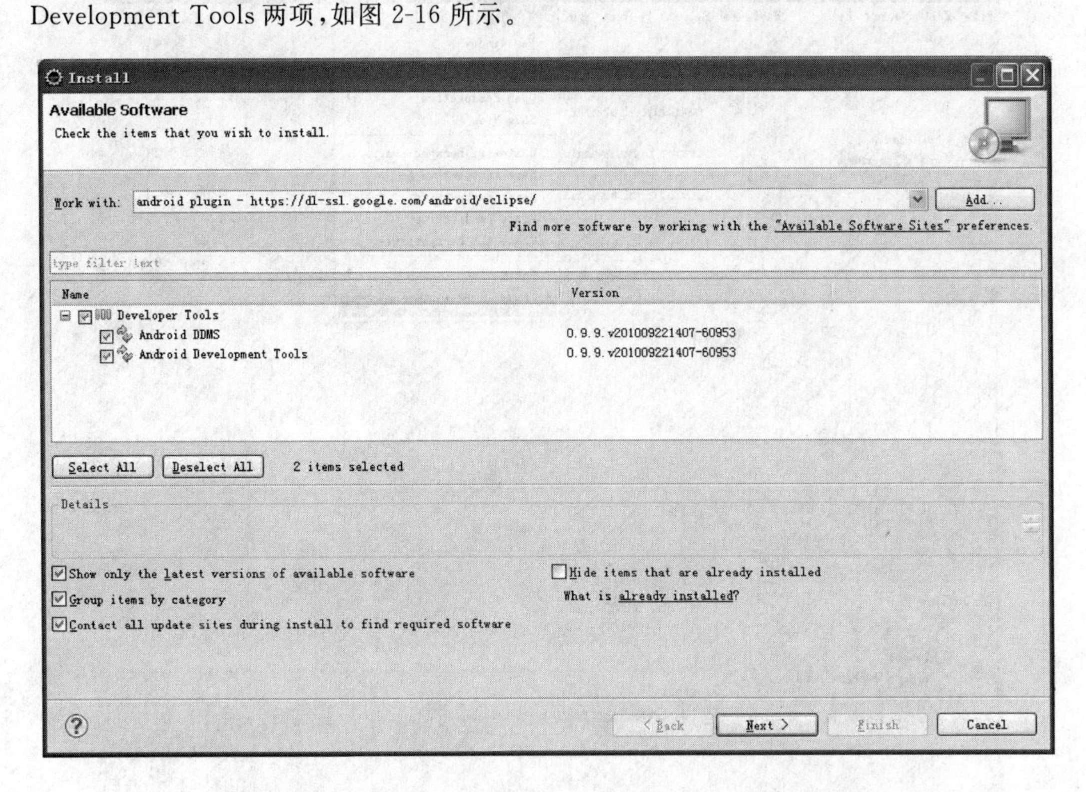

图 2-16　安装 Android DDMS 和 Android Development Tools

③ 接下来单击“Next”按钮，在下一个界面中选中接受协议复选项，最后单击“Finish”按钮，开始进行安装。安装成功后，需要重启 Eclipse，然后就可以使用 ADT 开发 Android 程序了。

(10) 配置 ADT 插件。

① 安装好插件后，还需要设置 Android SDK 的主目录才可以使用 Eclipse 创建 Android 工程。依次单击“Windows”→“Preferences”菜单选项，如图 2-17 所示。

② 在弹出的对话框左侧可以看到“Android”项，选中后在右侧设定 Android SDK 所在的目录为 SDK Location，单击“OK”按钮完成安装，如图 2-18 所示。

③ 设置好 Android SDK 的主目录后即可用 Eclipse 创建 Android 工程了。

(11) 安装模拟器。

① Android 工程需要发布到 Android 模拟器，因此需要创建一个虚拟设备。首先必须设置虚拟设备的名称、模拟器的版本、SD 卡的大小（这只是一个虚拟的 SD 卡）；然后，选择屏幕大小。具体如图 2-19 所示。

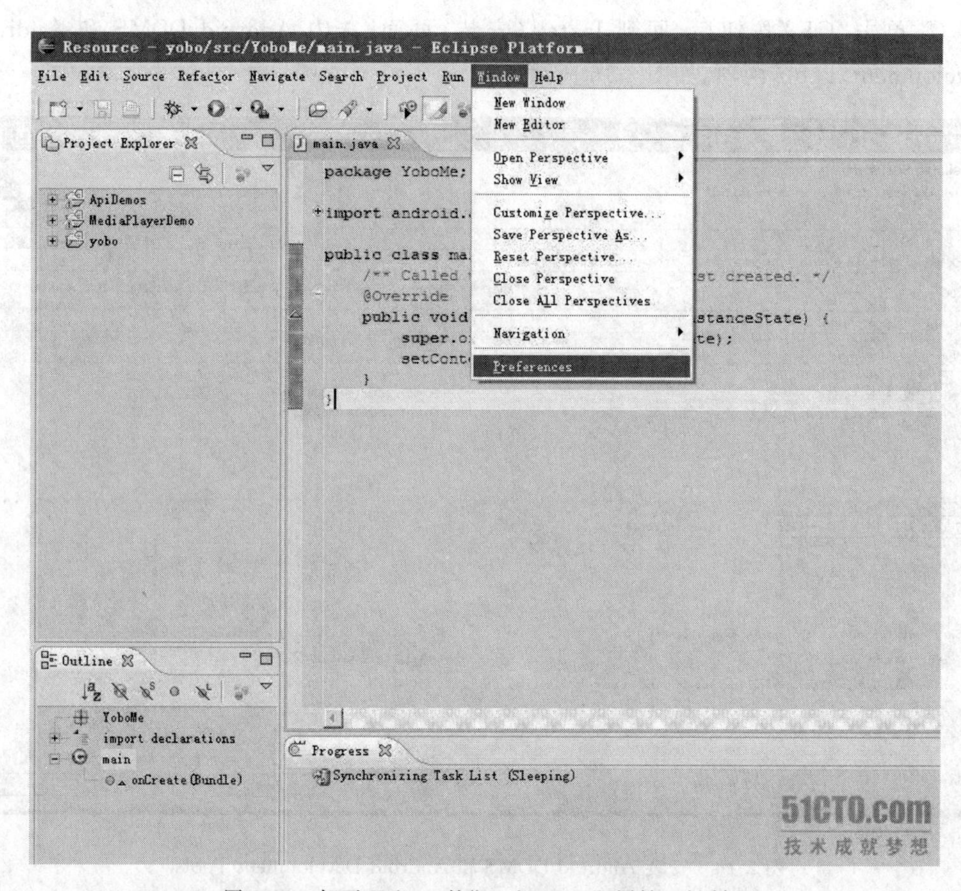

图 2-17　打开 Eclipse 的“Preferences”(属性)对话框

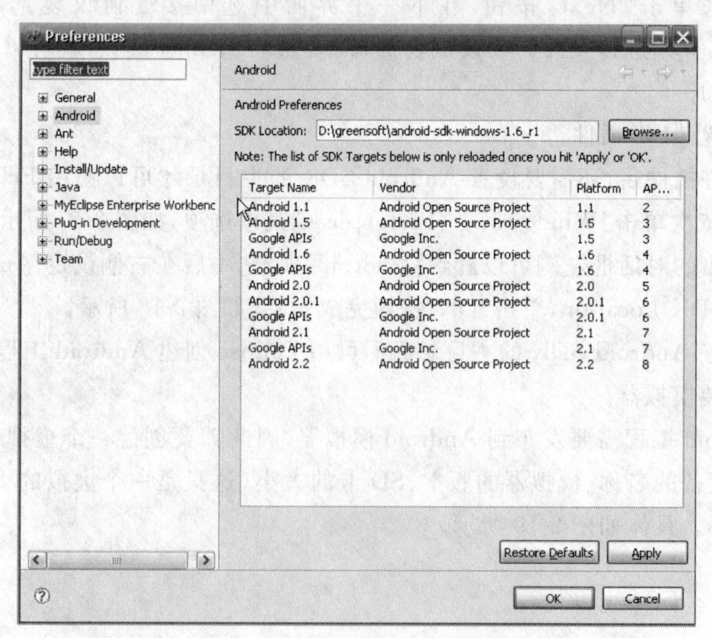

图 2-18　设置 Android SDK 的主目录

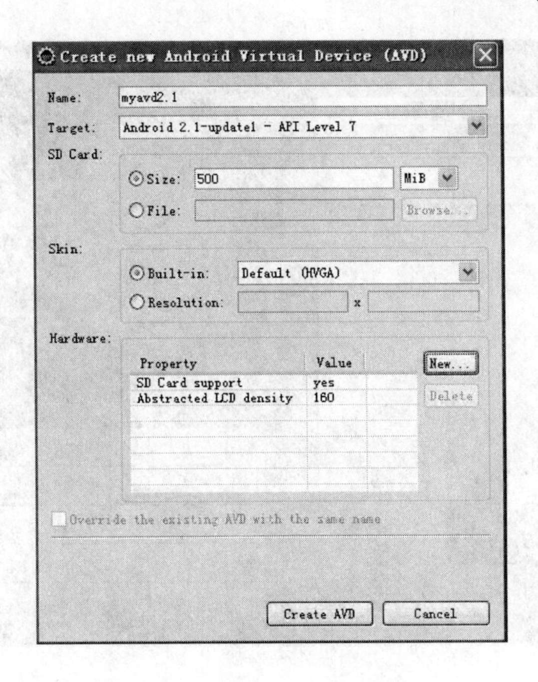

图 2-19 创建 Android 模拟器

② 现在运行 Android 模拟器。首先选择创建的虚拟设备，并单击右侧的“Start”按钮，如图 2-20 所示。

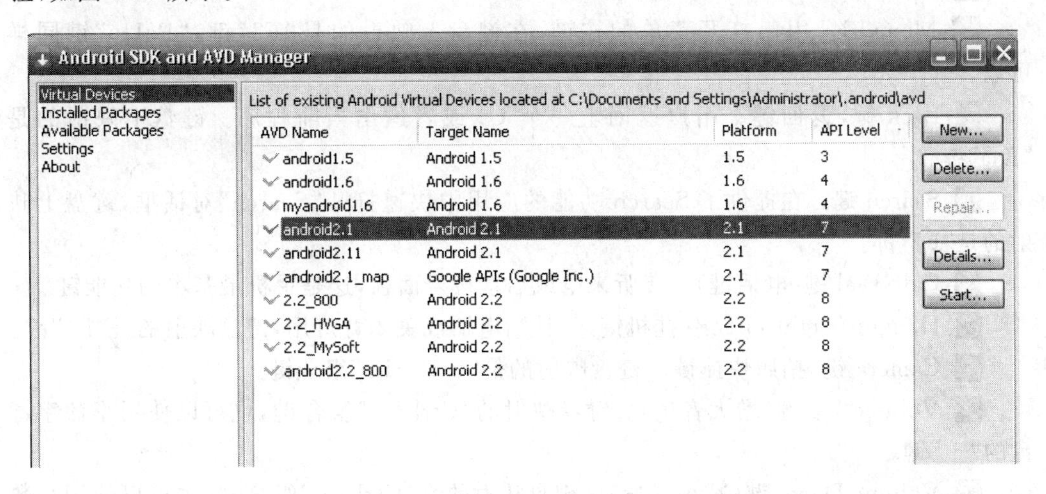

图 2-20 虚拟设备管理器

③ 随后，模拟器开始加载 Android 。可能会先打开几个命令提示符窗口，然后就可以看到模拟器本身。注意，在默认情况下，模拟器的右边会显示虚拟的按钮及键盘，如图 2-21所示。

图 2-21 Android 模拟器

模拟器中的主要按键及其功能介绍如下。

:Home 键。单击后直接显示桌面。

:Menu 键。用于打开菜单的按键,在键盘上映射的是“F2”键,“PgUp”键同样可以。

:Back 键,返回键。用户返回上一个 UI 或者退出当前程序。键盘上映射的是“Esc”键。

:Search 键。在提供了 Search 功能的应用中快速打开“Search”对话框,键盘上映射的是“F5”键。

:Call/Dial 键(电话键)。接听来电或启动拨号面板,这是手机最基本的功能键。

:Hangup/Light Off 键(挂机键)。挂断电话或关闭背灯用,键盘映射的是“F4”键。

:Camera 键,拍照快捷键。键盘映射的是“Ctrl+F3”组合键。

:Volume Up 键(增大音量)。键盘映射的“Ctrl+5”组合键,也可以使用小数字键盘的“+”键。

:Volume Down 键(减小音量)。键盘映射的是“Ctrl+6”组合键,也可以使用小数字键盘的“-”键。

:Power Down 键(关闭电源)。对应模拟器左上边缘的电源按钮,键盘映射的是“F7”键。

2.3.3 开发环境测试

开发 Android 应用时,一般是在模拟器环境中进行调试,但最终一定要在真机上进行测试。真机与模拟器还是有很大的区别的,下面介绍联机调试的基本步骤:

(1) 下载手机驱动并安装，保证手机可以与PC正常连接，也可以借助第三方软件(91助手)来完成此任务。

(2) 将手机状态改为"调试模式"。打开手机，在应用程序列表中找到"设置"，然后依次选中"设置"→"应用程序设置"→"开发"→"USB调试"选项，然后用手机连接PC。

(3) 打开IDE环境(Eclipse)的DDMS，此时能看到手机设备。

(4) 在开发工具的Run Configuration或Debug Configuration中配置Target为Manual。

(5) 在AndroidManifest.xml中的application标签中添加"android:debuggable="true""。

通过以上步骤可直接把程序联机到真机运行和测试。

第3章 Android应用程序基础

本章介绍 Android 应用程序的基础知识及其构成，包括四大组件 Activity、Service、BroadcastReceiver、ContentProvider 以及 Intent，其中重点介绍 Activity 和 Intent 的创建与使用。

3.1 应用程序基础

3.1.1 应用程序的组成

Android 应用程序没有一个单一的入口点（例如，没有 main()函数）。相反，系统要实例化和运行需要几个必要的组件。本节将介绍 Android 支持的 4 个重要组件，并不是每一个 Android 应用程序都同时需要这 4 个组件，某些时候，只需要几种来组合成一个应用。本章将对这些组件的功能进行简单的介绍，本书的其他章节将详细讲解各个组件的使用方法。

(1) 活动(Activity)组件

Activity 是 Android 构造块中最基本的一个组成单元，一般情况下，一个 Activity(活动)就是一个单独的屏幕。一个活动表示一个可视化的用户界面，关注用户的事件触发。每个活动都是独立于其他活动的，而且是作为 Activity 基类的一个子类来实现的。

(2) 服务(Service)组件

服务没有可视化的用户界面，而是在后台无期限地运行。每个服务都继承自 Service 基类。

(3) 广播接收者(Broadcast Receiver)组件

一个广播接收者就是一个组件，它仅仅只是接收广播的公告并对其作出相应的反应。许多广播是由系统发出的，一个应用程序可以有任意数量的广播接收者去响应任何它认为重要的公告。所有的接收者继承自 BroadcastReceiver 基类。

(4) 内容提供者(Content Provider)组件

Content Provider 是数据的包装器。它将一个应用程序指定的数据集提供给其他应用程序使用。这些数据集可能来自 XML 文件，也可能来自 SQLite 数据库，还可能来自网络服务器。内容提供者继承自 ContentProvider 基类，其设计了一些标准的方法集，使得其他应用程序可以检索和存储数据。

3.1.2 应用程序开发目录结构

在第 2 章的 HelloAndroid 示例中，我们已经看到了 Android 工程的目录结构，现在，结合图 3-1 再详细来讲解各个目录的作用。

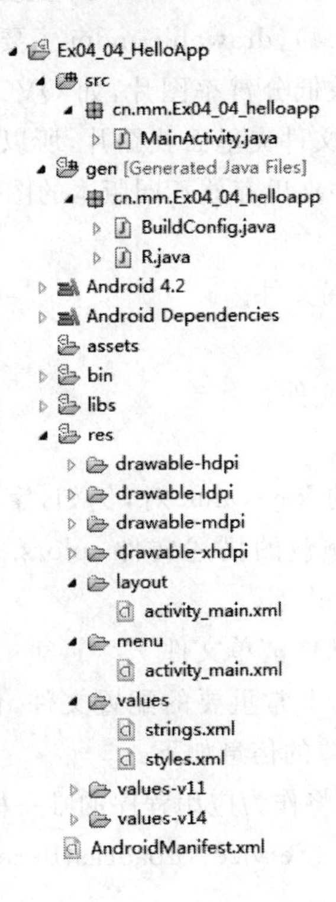

图 3-1 Android 工程的目录结构

(1) src：源码目录

src 目录中存放的是该项目的源代码，其内部结构会根据用户所声明的包自动组织。例如，在图 3-1 所示的目录结构中，该目录的组织方式为 src/cn/mm/Ex04_04_HelloApp. java，在项目开发过程中，程序员的大部分时间都在对该目录下的源代码文件进行编写。

(2) gen：R 文件目录

gen 目录中存放的是由 Android 开发工具自动生成的文件 R. java。R. java 是目录下唯一的文件，它的作用是管理程序中的资源(Res 目录下的资源)。这个目录中的内容是自动生成的，在开发过程中不需要开发人员去编辑。

(3) assets：放置非系统管理资源

该目录用于存放项目相关的数据或资源文件，例如文本文件、二进制文件等，这些文件不受系统管理。在程序中使用 getResources. getAssets(). open("text. txt")来得到资源文件的输入流，然后在程序中就可以读取其中的数据。

(4) res:放置各种资源文件

• res\drawable

该目录用于放置各种格式的图片文件，如 png、gif、jpg 等。Android SDK 从 1.5 版本以后，为了支持多分辨率的图片显示，res\drawable 分为 3 个目录 res\drawable-hdpi、res\drawable-ldpi、res\drawable-mdpi。drawable-hdpi 主要放高分辨率图片，如 WVGA (480×800)和 FWVGA (480×854)；drawable-mdpi 主要放中等分辨率图片，如 HVGA (320×480)；drawable-ldpi 主要放低分辨率图片，如 QVGA (240×320)。Android 系统根据机器的分辨率分别到对应的文件夹里面找图片，所以在开发程序时为了兼容不同平台的不同屏幕，建议在相应的文件夹里存放不同版本的图片。

• res\layout

该目录下存放的是 XML 布局文件。

• res\menu

该目录下存放的是菜单配置文件。

• res\values

该目录存放的是不同类型的 key-value 对，例如，字符串资源的描述文件 strings.xml，样式描述文件 styles.xml，颜色的描述文件 colors.xml，尺寸的描述文件 dimens.xml，以及数组描述文件 arrays.xml 等。

(5) AndroidManifest.xml:功能清单文件

AndroidManifest.xml 是一个非常重要的配置文件，包含应用程序中的每个组件的配置信息和权限信息等。可以配置的信息如下：

- 应用程序的包名。该包名将作为应用程序的唯一标识符。
- 所包含的组件，如 Activity、Service、BroadcastReceiver、ContentProvider 等。
- 应用程序兼容的最低版本。
- 声明应用程序需要的链接库。
- 应用程序自身应该具有的权限的声明。
- 其他应用程序访问应用程序时应该具有的权限。

(6) default.properties:项目环境信息

此文件用于存储开发环境的配置信息，开发人员一般不需要修改此文件。

随着学习的深入，我们还会接触更多的目录，后续章节会根据需要再作进一步介绍。

3.2 Android 应用程序的构成

Android 应用程序可能由一个或多个组件组成，下面将详细介绍这些组件。

3.2.1 Activity

Activity 是 Android 的核心组件，它为用户操作提供可视化的用户界面，每一个

Activity提供一个可视化的区域。比如说，一个短消息应用程序可以包括一个用于显示发送联系人列表的 Activity、一个写短信的 Activity，以及翻阅以前的短信和改变设置的 Activity。尽管它们彼此独立但一起组成了一个完整短消息应用程序界面。每个用户界面都以 Activity 作为基类。

一个应用程序可以包含一个或多个 Activity，Activity 的作用和数目取决于应用程序的设计。一般情况下，总有一个 Activity 被标记为用户在应用程序启动时第一个看到。

每个 Activity 至少提供一个窗口，窗口显示的可视内容是由一系列视图构成的，这些视图均继承自 View 基类。每个视图控制着窗口中一块特定的矩形空间。父视图包含并组织它的子视图的布局。叶节点视图(位于视图层次最底端)在它们控制的矩形中进行绘制，并对用户的操作作出响应。所以，视图是 Activity 与用户进行交互的界面。比如，视图可以显示一个小图片，并在用户指点它的时候产生动作。Android 有很多既定的视图供用户直接使用，包括按钮、文本域、卷轴、菜单项、复选框等。对于这些需要显示的组件在窗口中的位置摆放，则是通过 XML 布局文件来指定的。

3.2.2　Broadcast Receiver

广播接收器(Broadcast Receiver)是一个专注于接收广播信息并对其作出处理的组件。很多广播源自于系统广播。比如，通知时区改变、电池电量低、拍摄了一张照片或者用户改变了语言选项等，系统都会发出广播。用户应用程序也可以发送广播。

应用程序可以拥有任意数量的广播接收器以对所有它感兴趣的通知信息予以响应。所有的接收器均继承自 BroadcastReceiver 基类。

广播接收器不需要用户界面，它们可以启动一个 Activity 来响应收到的信息。

3.2.3　Service

服务没有可视化的用户界面，而是在后台运行。比如一个服务可以在用户做其他事情的时候在后台播放背景音乐，可以从网络上获取一些数据或者计算一些东西并提供给需要这个运算结果的 Activity 使用。每个服务都继承自 Service 基类。

一个媒体播放器播应用程序可能有一个或多个 Activity 来给用户选择歌曲并进行播放。然而，音乐播放这个任务本身不应该为任何 Activity 所处理，因为用户期望在他们离开播放器应用程序时，音乐仍在继续播放。为达到这个目的，媒体播放器 Activity 应该启用一个运行于后台的服务。

开发者可以连接(绑定)到一个正在运行的服务(如果服务没有运行，则启动它)。连接后，可以通过服务暴露出来的接口与服务进行通信。对于音乐服务来说，这个接口可以允许用户暂停、回退、停止以及重新开始播放。

服务运行于应用程序进程的主线程内，如果有 Activity 存在，它们将在同一主线程中。所以它可能会对其他组件或用户界面有干扰，因此在进行一些耗时的任务(比如音乐回放)时，一般会派生一个新线程来完成这个任务。

3.2.4 Content Provider

内容提供者将一些特定的应用程序数据提供给其他应用程序使用。数据可以存储于文件系统、SQLite 数据库或其他方式。内容提供者继承自 ContentProvider 基类，为其他应用程序读取和存储它管理的数据实现了一套标准方法。然而，应用程序并不直接调用这些方法，而是使用一个 ContentResolver 对象，调用它的方法来实现对 ContentProvider 数据的访问。ContentResolver 可以与任何内容提供者进行会话，从而对所有交互通信进行管理。

3.2.5 Intent

接收到 ContentResolver 发出的请求后，内容提供者即被激活，而其他 3 种组件（活动、服务和广播接收者）则被一种叫做意图（Intent）的异步消息激活。意图是一个保存着消息内容的 Intent 对象。对于活动和服务来说，Intent 对象指明了请求的操作名称以及作为操作对象的数据 URI 和其他一些信息。例如，Intent 可以传递一个对 Activity 的请求，让它为用户显示一张图片，或者让用户编辑一些文本。而对于广播接收者而言，Intent对象指明了广播的行为。对于每种组件来说，激活的方法是不同的。下面我们将首先了解 Activity 的使用以及如何用 Intent 激活它。

3.3 Activity 与 Intent

3.3.1 Activity 生命周期

应用程序组件都有生命周期，它们由 Android 初始化，直到这些实例被销毁。本小节讨论 Activity 的生命周期，包括它们在生命周期中的状态、在状态之间转变时的方法，以及当进程被关闭或实例被销毁时这些状态产生的效果。

一个 Activity 主要有 3 个状态：

(1) 当在屏幕前台时（位于当前任务栈的顶部），它处于运行的状态，即用户当前操作的 Activity。

(2) 当它失去焦点但仍然对用户可见时，它处于暂停状态，即在它之上有另外一个 Activity。这个 Activity 也许是透明的，或者未能完全遮蔽全屏，所以被暂停的 Activity 仍对用户可见。暂停的 Activity 仍然处于存活状态（它保留着所有的状态和成员信息并连接至窗口管理器），但当系统处于极低内存的情况下，系统仍然可以杀死这个 Activity。

(3) 当它完全被另一个 Activity 覆盖时，它处于停止状态，它仍然保留所有的状态和成员信息。它不再对用户可见，所以它的窗口将被隐藏，如果其他地方需要内存，则系统经常会杀死这种状态的 Activity。

如果一个 Activity 处于暂停或停止状态，系统可以要求它结束（调用它的 finish() 方法）或直接杀死它的进程来将它驱逐出内存。当它再次对用户可见的时候，它只能完全重

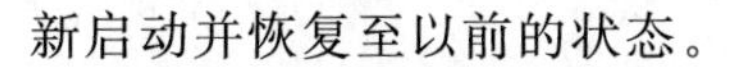

新启动并恢复至以前的状态。

当一个 Activity 从这个状态转变到另一个状态时，它被下列 protected 方法所通知：

(1) void onCreate(Bundle savedInstanceState)；

(2) void onStart()；

(3) void onRestart()；

(4) void onResume()；

(5) void onPause()；

(6) void onStop()；

(7) void onDestroy()。

开发者可以重载这些方法以在状态改变时执行合适的操作。所有的 Activity 都必须实现 onCreate()，便于对象在第一次实例化时进行初始化设置。很多 Activity 会实现 onPause()方法，主要是在页面发生变化时，能将重要数据持久保存到应用程序的数据存储中。

所有 Activity 生命周期方法的实现都必须先调用其父类的重写方法。

总的来说，这 7 个方法定义了一个 Activity 的完整生命周期。要实现这些方法可以查看 3 个嵌套的生命周期循环：

• 一个 Activity 的完整生命周期自第一次调用 onCreate()开始，直至调用 onDestroy()结束。Activity 在 onCreate()中设置所有"全局"状态以完成初始化，而在 onDestroy()中释放所有系统资源。比如，如果 Activity 有一个线程在后台运行，然后从网络上下载数据，它会用 onCreate()创建那个线程，而用 onDestroy()来销毁那个线程。

• 一个 Activity 的可视生命周期自 onStart() 调用开始，直到相应的 onStop()调用结束。在此期间，用户可以在屏幕上看到此 Activity。可以通过这两个方法来管控向用户显示这个 Activity 的资源。比如，可以在 onStart() 中注册一个 BroadcastReceiver 来监控会影响到用户改变的动作，在 onStop() 中来取消注册这个广播。onStart() 和 onStop() 方法可以随着应用程序是否对用户可见而被多次调用。

• 一个 Activity 的前台生命周期自 onResume()调用起，至相应的 onPause()调用结束。在此期间，Activity 位于前台的最上面并与用户进行交互。Activity 经常在暂停和恢复之间进行转换。比如说，当设备转入休眠状态或有新的 Activity 启动时，将调用 onPause() 方法。当 Activity 获得结果或者接收到新的 Intent 的时候，会调用 onResume()。因此，这两个方法中的代码应当是轻量级的。

图 3-2 展示了上述循环过程以及 Activity 在这个过程之中历经的状态改变。

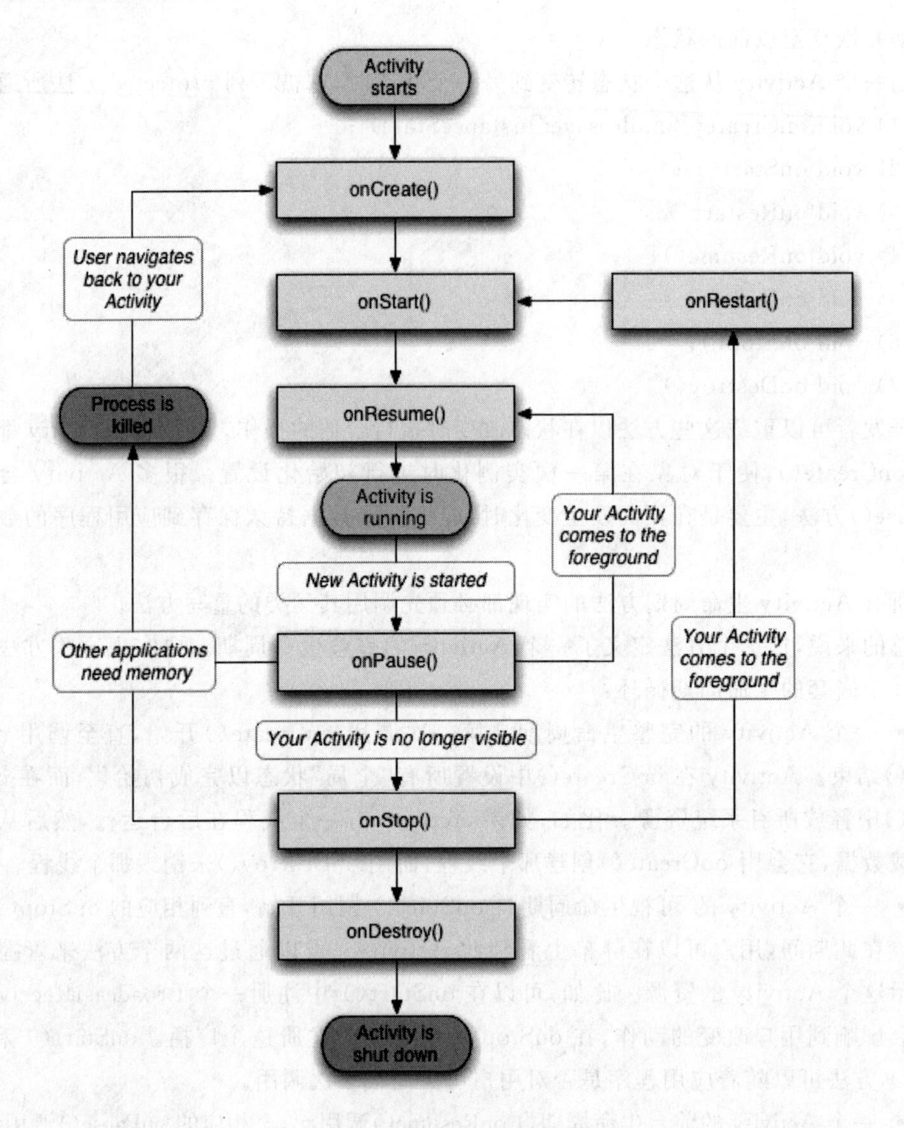

图 3-2　Activity 的生命周期

3.3.2　创建 Activity

前面阐述了 Activity 的生命周期及其原理，接下来通过代码来验证学过的知识。首先学会创建 Activtiy，在创建之前，先创建一个 Android 工程，然后按以下步骤完成创建。

(1) 创建一个自定义 Activtiy 类，此类继承自 Activity。

```
public class MainActivity extends Activity {
}
```

(2) 覆盖 onCreate(Bundle savedInstanceState)方法，当 Activity 第一次运行时，Activity框架会调用这个方法。

```
public class MainActivity extends Activity {
    /** Activity被第一次加载时调用 */
```

```
    @Override
    public void onCreate(Bundle savedInstanceState) {
        super.onCreate(savedInstanceState);
        // 加载主界面
        setContentView(R.layout.main);
// 获取 UI 组件引用
        findViews();
// 为组件添加事件监听
        setListensers();
    }
}
```

(3) 在 AndroidManifest 文件中配置 Activtiy 信息。由于 Activity 是 Android 应用程序的一个组件，所以每一个 Activity 都需要在 Android 的配置文件中进行配置。

```
<? xml version = "1.0" encoding = "utf-8"? >
    <manifest
    xmlns:android = "http://schemas.android.com/apk/res/android"
    package = "cn.mm.Ex05"
    android:versionCode = "1"
    android:versionName = "1.0">
<application android:icon =
"@drawable/icon" android:label = "@string/app_name">
<activity android:name =
".MainActivity" android:label = "@string/app_name">
<intent-filter>
<action android:name = "android.intent.action.MAIN" />
    <category android:name = "android.intent.category.LAUNCHER" />
</intent-filter>
</activity>
</application>
</manifest>
```

(4) 为 Activity 添加必要的控件，在 layout 文件夹中创建一个 xml 格式的布局文件，然后再在这个布局文件中对 Activity 的布局以及不同的控件进行设置。

```
<? xml version = "1.0" encoding = "utf-8"? >
    <LinearLayout xmlns:android = "http://schemas.android.com/apk/res/android"
      android:orientation = "vertical" android:layout_width = "fill_parent"
  android:layout_height = "fill_parent">
  <TextView android:id = "@ + id/testMessage"
      android:layout_width = "fill_parent"
      android:layout_height = "wrap_content"
      android:text = "@string/hello" />
  <Button
```

```
            android:id="@+id/testButton"
            android:layout_width="wrap_content"
            android:layout_height="wrap_content"
            android:text="change" />
</LinearLayout>
```

(5) 在第一步定义的 Activity 子类中通过 findViewById()方法来获取布局文件中声明的控件,前提是布局文件中必须声明这些控件的 id。

```
//私有方法,查找组件
    private TextView testMessage;
    private Button testButton;
    private void findViews(){
  testMessage=(TextView)findViewById(R.id.testMessage);
  testButton=(Button)findViewById(R.id.testButton);
}
```

(6) 在相应的组件上添加事件处理。

```
//匿名内部监听类
    private Button.OnClickListener calcBMI =
new Button.OnClickListener()
{
    public void onClick(View v){
        testMessage.setText("Changed OK!");
    }
        };

        //私有方法,注册监听
        private void setListensers(){
    testButton.setOnClickListener(calcBMI);
    }
```

完成上述步骤后,可以利用第 2 章的知识来发布应用程序到真机或模拟器上,以便进行相关测试。

3.3.3 使用 Intent 跳转 Activity

在上一小节中,我们学会了如何创建一个单独的 Activity,但是一个应用程序应该由若干 Activity 组成,接下来,我们看看如何在这些 Activity 之间实现跳转并传递数据。

1. 用类名跳转

Intent 对象对操作的动作、动作涉及的数据、附加信息进行描述,Android 首先根据此 Intent 的描述找到对应的组件,然后将 Intent 传递给调用的组件,并完成组件调用。Intent 实现了调用者与被调用者之间的解耦作用。Intent 在传递过程中要找到目标消费者,也就是 Intent 的响应者。

下面我们看看用类名直接跳转的代码:

```
Intent intent = new Intent();              //构建意图对象
    //在 Intent 上设置来源的 Activity 实例,和要前往的 Activity 所在的 class
intent.setClass(A.this,B.class);
startActivity(intent);                     //开始跳转
```

2. 用 Action 来跳转

(1) 自定义 Action

如果在 AndroidManifest.xml 中有一个 Activity 的 IntentFilter 定义了一个 Action,那么请求的 Intent 能与这个目标 Action 匹配,并且 IntentFilter 中没有定义 Type,Category 内容,那么这个 Activity 就匹配了,此时系统就会启动这个 Activity。如果有两个以上的程序匹配,系统就会弹出一个对话框来提示说明。Action 的值在 Android 中有很多预定义,如果想直接转到自己定义的 Intent 接收者,可以在接收者的 IntentFilter 中加入一个自定义的 Action 值(同时要设定 Category 的值为"android.intent.category.DEFAULT"),在 Intent 中设定该值为 Intent 的 Action,就直接能跳转到自己的 Intent 接收者中。

下面是在 AndroidManifest.xml 文件中自定义的 Action 动作:

```
<!-- 配置跳转 activity -->
<activity android:name="com.android.dialog.myDialog">
<intent-filter>
<!-- 配置 action 路径 -->
    <action android:name="android.intent.action.mydialog" />
    <category android:name="android.intent.category.DEFAULT" />
</intent-filter>
</activity>
```

(2) 系统 Action

接下来,列举了很多系统 Action,并传递了不同的 Data 来激活应用程序以外的 Activity。这些代码大家可以在以后的程序中直接使用,比如显示一个网址、打电话、发短信、显示地图等。

```
//显示网页
   Uri uri = Uri.parse("http://google.com");
  Intent it = new Intent(Intent.ACTION_VIEW, uri);
  startActivity(it);
//显示地图
  Uri uri = Uri.parse("geo:38.899533,-77.036476");
  Intent it = new Intent(Intent.ACTION_VIEW, uri);
  startActivity(it);
//打电话
  //叫出拨号程序
  Uri uri = Uri.parse("tel:0800000123");
  Intent it = new Intent(Intent.ACTION_DIAL, uri);
  startActivity(it);
```

```
    //直接打电话出去
    Uri uri = Uri.parse("tel:0800000123");
    Intent it = new Intent(Intent.ACTION_CALL, uri);
    startActivity(it);
    //用这个,要在 AndroidManifest.xml 中加上
    //<uses-permission id = "android.permission.CALL_PHONE" />
//传送 SMS/MMS
    //调用短信程序
    Intent it = new Intent(Intent.ACTION_VIEW, uri);
    it.putExtra("sms_body", "The SMS text");
    it.setType("vnd.android-dir/mms-sms");
    startActivity(it);
    //传送消息
    Uri uri = Uri.parse("smsto://0800000123");
    Intent it = new Intent(Intent.ACTION_SENDTO, uri);
    it.putExtra("sms_body", "The SMS text");
    startActivity(it);
    //传送 MMS
    Uri uri = Uri.parse("content://media/external/images/media/23");
    Intent it = new Intent(Intent.ACTION_SEND);
    it.putExtra("sms_body", "some text");
    it.putExtra(Intent.EXTRA_STREAM, uri);
    it.setType("image/png");
    startActivity(it);
// 传送 E-mail
    Uri uri = Uri.parse("mailto:xxx@abc.com");
    Intent it = new Intent(Intent.ACTION_SENDTO, uri);
    startActivity(it);
    Intent it = new Intent(Intent.ACTION_SEND);
    it.putExtra(Intent.EXTRA_EMAIL, "me@abc.com");
    it.putExtra(Intent.EXTRA_TEXT, "The email body text");
    it.setType("text/plain");
    startActivity(Intent.createChooser(it, "Choose Email Client"));
    Intent it = new Intent(Intent.ACTION_SEND);
    String[] tos = {"me@abc.com"};
    String[] ccs = {"you@abc.com"};
    it.putExtra(Intent.EXTRA_EMAIL, tos);
    it.putExtra(Intent.EXTRA_CC, ccs);
    it.putExtra(Intent.EXTRA_TEXT, "The email body text");
    it.putExtra(Intent.EXTRA_SUBJECT, "The email subject text");
    it.setType("message/rfc822");
```

```
    startActivity(Intent.createChooser(it, "Choose Email Client"));
    //传送附件
    Intent it = new Intent(Intent.ACTION_SEND);
    it.putExtra(Intent.EXTRA_SUBJECT, "The email subject text");
    it.putExtra(Intent.EXTRA_STREAM, "file:///sdcard/mysong.mp3");
    sendIntent.setType("audio/mp3");
    startActivity(Intent.createChooser(it, "Choose Email Client"));
//播放多媒体
        Uri uri = Uri.parse("file:///sdcard/song.mp3");
        Intent it = new Intent(Intent.ACTION_VIEW, uri);
        it.setType("audio/mp3");
        startActivity(it);
        Uri uri =
Uri.withAppendedPath(MediaStore.Audio.Media.INTERNAL_CONTENT_URI,
            "1");
        Intent it = new Intent(Intent.ACTION_VIEW, uri);
        startActivity(it);
```

3. 传递数据

使用 Intent 在页面之间跳转，数据传递是必须的，我们可以直接在 Intent 对象上放置基本数据类型的数据，也可以放置字符串和其他数据类型数据。对于其他数据类型，实现了 Parcelable 或 Serializable 接口就可以。

下面的代码是用 Bundle 来实现数据传递，请参考：

```
      Intent intent = new Intent();                    //意图
      //在 Intent 上设置来源的 Activity 的实例和要前往的 Activity 所在的 class
intent.setClass(A.this,B.class);
   //附加在 Intent 上的消息都存储在 Bundle 实体对象中(map 对象)
Bundle bundle = new Bundle();
   bundle.putString("KEY_HEIGHT","200");             //附带信息
   bundle.putString("KEY_WEIGHT","300");
//将 Bundle 对象附加在 Intent 对象上
   intent.putExtras(bundle);
   startActivity(intent);                           //跳转
```

基本UI设计

UI 即 User Interface(用户界面)的简称。UI 设计是指对应用软件的人机交互、操作逻辑、界面美观的整体设计。好的 UI 设计不仅让应用软件变得有个性有品味,还让操作变得舒适、简单、自由,增强用户对软件定位和特点的体验。

本章介绍 Android 平台应用程序的基本 UI 组件(包括 TextView、EditText、Button、ImageButton、ImageView、RadioButton)和五大布局方式(包括 LinearLayout、FremeLayout、TableLayout、AbsoultLayout、RelativeLayout),并介绍 Andriod 平台的事件处理。通过本章的学习,将初步掌握 UI 设计的基本概念和方法,为进一步学习高级 UI 设计奠定基础。

4.1 基本 UI 组件

4.1.1 TextView 类

我们创建的第一个工程“HelloAndroid”就是用 TextView 来显示一段文字。TextView 是一个用来显示文本标签的组件,下面我们把“HelloAndriod”的实现代码改写一下,已修改 TextView 显示的文字的颜色、大小等属性,运行效果如图 4-1 所示。

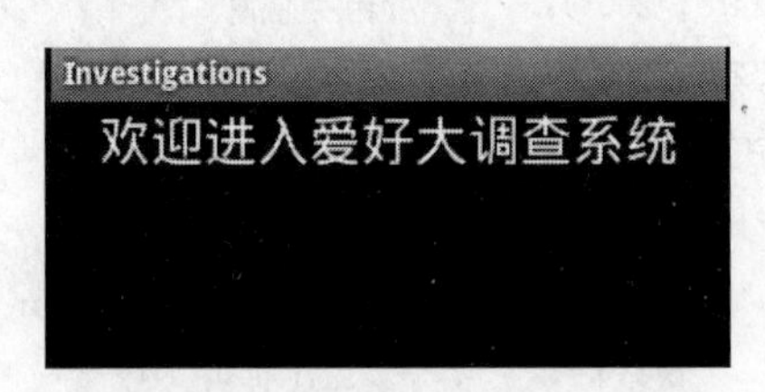

图 4-1 TextView 效果图

首先,我们来看一下在 Ex06_02_01_Investigations/res/layout/main.xml 布局文件中 TextView 的定义:

```
<TextView
    android:layout_width="fill_parent"
    android:layout_height="wrap_content"
    android:text="@string/hello"
```

```
android:gravity="center"android:textColor="#ffffffff"
android:textSize="25px"android:id="@+id/textViewHello"/>
```

在TextView标签中,android:id属性代表了TextView组件的id值;android:layout_width属性指定了组件的基本宽度;android:layout_height属性指定了组件的基本高度,一般只能设置为fill_parent(填充整个屏幕)或wrap_content(填充组件内容本身大小);android:text属性表示TextView显示的文字内容;android:textColor属性设置了TextView显示的文字的颜色,需要注意的是,颜色值只能是以“#”开头的8位0~f之间的值,前两位代表的是透明度,后6位是颜色值;android:textSize属性设置了TextView显示的文字的字体大小;android:gravity属性是对该view内容的限定,这里是将TextView里的文字居中显示。通过setText()方法可以修改TextView显示的文字。

如果我们的TextView对象里是一个URL地址,而且需要以链接的形式显示的时候,我们可以对TextView执行以下操作,在布局文件里为TextView加上android:autoLink=“all”属性,“all”是匹配所有的链接,具体如下面的代码所示。

```
<TextView
  android:id="@+id/tv"
    android:layout_width="fill_parent"
    android:layout_height="wrap_content"
    android:autoLink="all"
    android:text="有问题问Google:http://www.google.hk"
/>
```

从上边的代码中可以看出TextView不仅能显示文本而且还能识别文本中的链接,并将这些链接转换成可单击的链接。系统会根据不同类型的链接调用相应的软件进行处理,当我们单击上面例子里的链接时,系统会启动Android内置的浏览器,并导航到网址指向的网页。

注意:我们后面要学习的每个组件都有很多属性,具体的属性内容可以查看Android的官方文档。

4.1.2 EditText类

EditText也是开发中经常使用的组件,比如,要实现一个登录界面,需要用户输入账号和密码等信息,然后我们获得用户输入的内容,把它交给服务器来判断。因此,我们要学习如何在布局文件中实现编辑框,然后获得编辑框的内容。

程序运行效果如图4-2所示。

图4-2 EditText效果图

EditText在Ex06_02_01_Investigations/res/layout/main.xml布局文件中的定义如下所示:

```
<EditTextandroid:layout_height="wrap_content"
android:layout_width="fill_parent"android:id="@+id/editTextName"
android:hint="请输入名字">
</EditText>
```

在 EditText 标签中，android:hint 属性是设置 EditText 为空时输入框内的提示信息。

Ex06_02_01_Investigations/src/cn/mm/Ex06_02_01_Investigations/MainActivity 中的代码片段如下：

```
private EditText nameEditText = null; //声明需要用到的组件
@Override
publicvoid onCreate(Bundle savedInstanceState)
{
super.onCreate(savedInstanceState);
//将 main.xml 布局文件加载到 MainActivity 中
        setContentView(R.layout.main);
//通过 findViewById 方法找到 EditText 组件,强制类型转换
nameEditText = (EditText) findViewById(R.id.editTextName);
}
```

我们在 Activity 类中通过使用 findViewById()方法获得 EditText 对象，获得 EditText 编辑框中的内容的是 getText()方法。

4.1.3 Button 类

按钮(Button)是用得最多的组件之一，在前面的例子中，我们已经使用到了按钮。既然是按钮，必然有按钮的触发事件，所以需要通过 setOnClickListener 事件来监听。在下面的例子中，我们通过单击按钮触发了一个跳转事件。

程序运行效果如图 4-3 所示。

图 4-3 Button 效果图

Button 在 Ex06_02_01_Investigations/res/layout/main.xml 布局文件中的定义如下所示：

```
<Button>
    android:layout_width="wrap_content"
    android:layout_height="wrap_content"
    android:text="进入调查系统"android:id="@+id/buttonSubmit">
</Button>
```

我们可以设置按钮的大小、文本的颜色等属性，按钮的其他属性也是可以设置的。

Ex06_02_01_Investigations/src/cn/mm/Ex06_02_01_Investigations/MainActivity 中的代码片段如下所示：

```
private Button submitButton = null; //声明 Button 组件
//通过 findViewById 方法找到 Button 组件,强制类型转换
submitButton = (Button) findViewById(R.id.buttonSubmit);
submitButton.setOnClickListener( new OnClickListener()
 {//给按钮注册监听事件
  publicvoid onClick(View v) {
            Intent intent = new Intent();//创建一个 Intent 对象
            //指定 Intent 要启动的类
            intent.setClass(MainActivity.this, Inquiry.class);
            startActivity(intent); //启动一个新的 Activity
            MainActivity.this.finish();//关闭当前的 Activity
  }
});
```

4.1.4 ImageButton 类

除了我们已经介绍的 Button,我们还可以制作带图标的 Button(按钮),这就是接下来介绍的 ImageButton 组件。

程序运行效果如图 4-4 所示。

图 4-4 ImageButton 效果图

ImageButton 在 Ex06_02_01_Investigations/res/layout/inquiry.xml 布局文件中的定义如下所示:

```
<ImageButtonandroid:src="@drawable/icon"
  android:layout_width="wrap_content"
    android:layout_height="wrap_content"
    android:id="@+id/imageButtonOcc"
    android:layout_margin="10dp"
></ImageButton>
 <ImageButtonandroid:src="@drawable/icon"
```

```
    android:layout_width="wrap_content"
    android:layout_height="wrap_content"
    android:id="@+id/imageButtonHobb"
    android:layout_margin="10dp"
></ImageButton>
 <ImageButtonandroid:src="@drawable/icon"
        android:layout_width="wrap_content"
        android:layout_height="wrap_content"
        android:id="@+id/imageButtonPicture"
        android:layout_margin="10dp"
></ImageButton>
```

在 ImageButton 标签中,我们通过 android:src 设置 ImageButton 显示的图标。

Ex06_02_01_Investigations/src/cn/mm/Ex06_02_01_Investigations/InquiryActivity 的代码片段如下所示:

```
  private ImageButton professionButton = null; //声明 ImageButton 组件
  private ImageButton hobbyButton = null; //声明 ImageButton 组件
  private ImageButton pictureButton = null;
  private Intent intent = new Intent();  //创建 Intent 对象
@Override
  protectedvoid onCreate(Bundle savedInstanceState) {
    super.onCreate(savedInstanceState);
    setContentView(R.layout.inquiry);
          //获得 Button 组件
          professionButton =
                    (ImageButton) findViewById(R.id.imageButtonOcc);
          hobbyButton =
                    (ImageButton) findViewById(R.id.imageButtonHobb);
          pictureButton =
                    (ImageButton) findViewById(R.id.imageButtonPicture);
          //给 ImageButton 组件注册监听事件
          professionButton.setOnClickListener(new OnClickListener() {
      @Override
    publicvoid onClick(View v) {
          //指定 Intent 要启动的类
          intent.setClass(InquiryActivity.this,
                    ProfessionActivity.class);
          startActivity(intent); //启动一个新的 Activity
          InquiryActivity.this.finish();//关闭当前的 Activity
    }
  });
          //给 ImageButton 组件注册监听事件
  hobbyButton.setOnClickListener(new OnClickListener() {
```

```
        publicvoid onClick(View v) {
                //指定 Intent 要启动的类
                intent.setClass(InquiryActivity.this,
                            HobbyActivity.class);
    //启动一个新的 Activity 指定 Intent 要启动的类
                startActivity(intent);
                InquiryActivity.this.finish();//关闭当前的 Activity
            }
        });
                //给 ImageButton 组件注册监听事件
    pictureButton.setOnClickListener(new OnClickListener() {
        publicvoid onClick(View v) {
                //指定 Intent 要启动的类
                intent.setClass(InquiryActivity.this,
                            PictureActivity.class);
                startActivity(intent); //启动一个新的 Activity
                InquiryActivity.this.finish();//关闭当前的 Activity
            }
        });
      }
```

4.1.5　ImageView 类

如果我们想把一张图片显示在屏幕上，就需要创建一个显示图片的对象。在 Android 中，这个对象是 ImageView，可以通过 setImageResource 方法来设置要显示的图片资源索引，还可以对图片执行一些其他的操作，比如设置它的 Alpha(透明度)值等。

程序运行效果如图 4-5 所示。

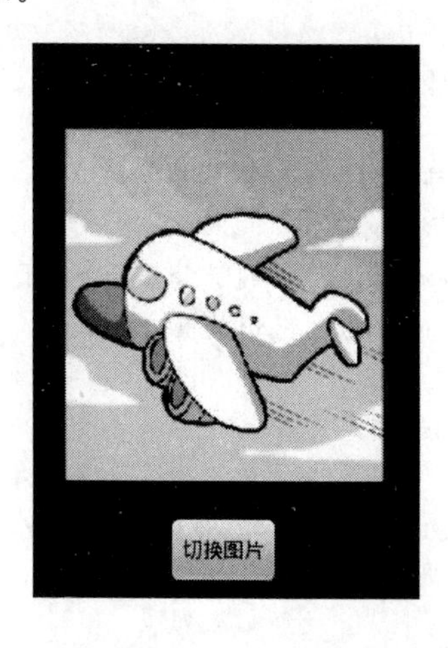

图 4-5　ImageView 效果图

ImageView 在 Ex06_02_01_Investigations/res/layout/picture.xml 布局文件中的定义如下所示：

```
<ImageView
  android:id="@+id/imageViewPic"
  android:layout_height="240dp"
  android:layout_width="240dp"
  android:layout_margin="10dp"android:src="@drawable/a"></ImageView>
<Buttonandroid:text="切换图片"android:id="@+id/buttonSure"
  android:layout_width="wrap_content"android:layout_height="wrap_content"
  android:layout_margin="10dp"></Button>
```

在 ImageView 标签里，我们通过设置 android:src 属性来显示图片；android:layout_margin 属性是图片离父容器上下左右 4 个方向的边缘的距离；android:layout_marginBottom 属性是图片离某元素底边缘的距离；android:layout_marginLeft 属性是图片离某元素左边缘的距离；android:layout_marginRight 属性是图片离某元素右边缘的距离；android:layout_marginTop 属性是图片离某元素上边缘的距离。

Ex06_02_01_Investigations/src/cn/mm/Ex06_02_01_Investigations/ PictureActivity 的代码片段如下所示：

```
publicclass PictureActivity extends Activity {
  private ImageView picImg = null; //声明 ImageView 组件
  private Button sureButton = null; //声明 Button 组件
  privateintid = R.drawable.a;  //定义 id 获得图片 id
  @Override
  protectedvoid onCreate(Bundle savedInstanceState) {
    super.onCreate(savedInstanceState);
    setContentView(R.layout.picture); //加载布局
    //获得 ImageView 对象
    picImg = (ImageView) findViewById(R.id.imageViewPic);
    //获得 Button 对象
    sureButton = (Button) findViewById(R.id.buttonSure);
    //注册监听事件
    sureButton.setOnClickListener(new OnClickListener() {
        publicvoid onClick(View v) {
            if(id = = R.drawable.a) //判断 id 值是否等于 a
              id = R.drawable.b; //改变 id 的值,切换图片
            else {
              id = R.drawable.a;
            }
            picImg.setImageResource(id); //设置 ImageView 要显示的图片
        }
    });
  }
}
```

代码显示图片的方法是 setImageResource(int id)，设置 ImageView 的背景的方法是 set-

BackgroundResource(resid)。

4.1.6 RadioButton 类

单项选择也是开发中使用得很多的组件，在投票或调查类应用中我们会经常用到。在Android 里，可通过 RadioGroup 和 RadioButton 一起来实现一个单项选择效果。Android 平台上的单选按钮可通过 RadioButton 来实现，而 RadioGroup 则是将 RadioButton 集中管理。我们通过一个例子来说明如何实现，程序运行的效果如图 4-6 所示。

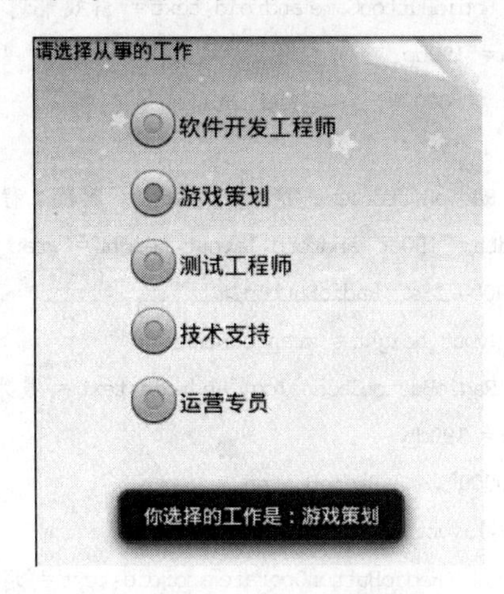

图 4-6　RadioButton 效果图

一个单项选择由两部分组成：分别是前面的选择按钮和后面所选择的“答案”。Android 平台上的选择按钮可通过 RadioButton 来实现，而选择的“答案”则通过 RadioGroup 来实现。因此，我们在布局文件中定义一个 RadioGroup 和 5 个 RadioButton，在定义 RadioGroup 时，已经将“答案”赋给了每个选项，那么如何确定用户的选择是否正确呢？这需要在用户执行单击操作时来判断用户所选择的是哪一项，需要为其注册事件监听的方法是 setOnCheckedChangeListener()。

首先，RadioButton 在 Ex06_02_01_Investigations/res/layout/profession.xml 布局文件中的定义如下所示：

```
<LinearLayoutxmlns:android="http://schemas.android.com/apk/res/android"
  android:layout_width="fill_parent"android:layout_height="fill_parent"
  android:orientation="vertical"android:background="@drawable/d"
ndroid:gravity="center">
<TextViewandroid:layout_height="wrap_content"
  android:layout_width="fill_parent"android:text="请选择从事的工作"
  android:textColor="#ff000000"
android:id="@+id/textViewDisplay"></TextView>
  <RadioGroupandroid:id="@+id/radioGroupPro"
      android:layout_height="wrap_content"
      android:layout_width="fill_parent"android:layout_margin="10dp"
```

```
        android:padding="10dp"android:gravity="center_horizontal"
        android:layout_weight="1">
    <RadioButtonandroid:text="软件开发工程师"
        android:layout_height="wrap_content"
        android:id="@+id/radioButtonSoft"android:layout_width="190dp"
        android:textColor="#ff000000"></RadioButton>
    <RadioButtonandroid:layout_height="wrap_content"
        android:id="@+id/RadioButtonGame"android:text="游戏策划"
        android:layout_width="190dp"
        android:textColor="#ff000000"></RadioButton>
      <RadioButton
        android:id="@+id/RadioButtonTest" android:text="测试工程师"
          android:layout_width="190dp" android:layout_height="wrap_content"
android:textColor="#ff000000"></RadioButton>
      <RadioButtonandroid:layout_height="wrap_content"
        android:id="@+id/RadioButtonTechnology"android:text="技术支持"
        android:layout_width="190dp"
android:textColor="#ff000000"></RadioButton>
        <RadioButtonandroid:layout_height="wrap_content"
            android:id="@+id/RadioButtonOperate"android:text="运营专员"
            android:layout_width="190dp"
android:textColor="#ff000000"></RadioButton>
</RadioGroup>
</LinearLayout>
```

Ex06_02_01_Investigations/src/cn/mm/Ex06_02_01_Investigations/ProfessionActivity 的代码片段如下所示：

```
//声明 RadionGroup 组件
private RadioGroup radioGroup=null;
//声明 RadionButton 组件
private RadioButton softRadioButton=null;
//声明 RadionButton 组件
private RadioButton testRadioButton=null;
//声明 RadionButton 组件
private RadioButton gameRadioButton=null;
//声明 RadionButton 组件
private RadioButton technologyRadioButton=null;
//声明 RadionButton 组件

private RadioButton operateRadioButton=null;
  @Override
protectedvoid onCreate(Bundle savedInstanceState) {
```

```
super.onCreate(savedInstanceState);
setContentView(R.layout.profession); //加载布局文件
//获得 RadionGroup 对象
radioGroup = (RadioGroup) findViewById(R.id.radioGroupPro);
//获得 RadioButton 对象
softRadioButton =
(RadioButton) findViewById(R.id.radioButtonSoft);
testRadioButton =
(RadioButton) findViewById(R.id.RadioButtonTest);
gameRadioButton =
(RadioButton) findViewById(R.id.RadioButtonGame);
technologyRadioButton =
(RadioButton) findViewById(R.id.RadioButtonTechnology);
operateRadioButton =
(RadioButton) findViewById(R.id.RadioButtonOperate);
radioGroup.setOnCheckedChangeListener(new  RadioGroup.OnCheckedChangeListener()
{        //注册监听事件
    @Override
publicvoid onCheckedChanged(RadioGroup group, int checkedId) {
  switch (checkedId) {//根据单选按钮 ID 显示用户所选择的信息
    case R.id.radioButtonSoft:
  showToast("你选择的工作是:" + softRadioButton.getText());
      break;
    case R.id.RadioButtonGame:
  showToast("你选择的工作是:" + gameRadioButton.getText());
break;
    case R.id.RadioButtonTest:
  showToast("你选择的工作是:" + testRadioButton.getText());
      break;
    case R.id.RadioButtonTechnology:
      showToast("你选择的工作是:" +
            technologyRadioButton.getText());
      break;
    case R.id.RadioButtonOperate:
  showToast("你选择的工作是:" + operateRadioButton.getText());
      break;
    }
  }
});
}
privatevoid showToast(String str){        //Toast 提示信息
  Toast.makeText(ProfessionActivity.this,
```

```
str, Toast.LENGTH_SHORT).show(); }
```

4.2 布局管理器

Android 提供了 5 种通用的布局对象,它们都是视图组的子类。它们分别是 FrameLayout(框架布局)、LinearLayout(线性布局)、TableLayout(表单布局)、AbsoluteLayout(绝对布局)和 RelativeLayout(相对布局)。下面我们将对这 5 种布局进行详细介绍。

4.2.1 FrameLayout

FrameLayout 是最简单的一个布局对象,所有的组件都会固定在屏幕的左上角,不能指定位置。一般而言,如果要制作一个复合型的新组件,都可以基于该类来实现。下面是 FrameLayout 布局代码。

FrameLayout 在 Ex06_03_Layout/res/layout/frame.xml 布局文件中的定义:

```
<? xml version = "1.0" encoding = "utf-8"? >
<FrameLayout xmlns:android = "http://schemas.android.com/apk/res/android"
    android:orientation = "vertical"
    android:layout_width = "fill_parent"
    android:layout_height = "fill_parent"
>
    <TextView
        android:layout_width = "fill_parent"
        android:layout_height = "wrap_content"
        android:text = "这里是文字 1"/>
    <EditText
        android:text = "这里是编辑文字"
        android:id = "@ + id/EditText01"
        android:layout_width = "wrap_content"
        android:layout_height = "wrap_content">
    </EditText>
    <Button
    android:text = "这里是按钮 1"
        android:id = "@ + id/Button01"
        android:layout_width = "wrap_content"
        android:layout_height = "wrap_content">
    </Button>
</FrameLayout>
```

在 xml 布局中,android:orientation 属性声明了 FrameLayout 布局的排列方式(水平还是垂直)。

FrameLayout 的布局效果如图 4-7 所示。

图 4-7 FrameLayout 效果图

4.2.2 LinearLayout

LinearLayout 的功能是以单一方向对其中的组件进行线性排列显示。比如,以垂直排列显示,则各组件将在垂直方向上排列显示;以水平排列显示,则各组件将在水平方向上排列显示。下面是关于 LinearLayout 的布局代码。

LinearLayout 在 Ex06_03_Layout/res/layout/ linear. xml 布局中的定义:

```
<?xml version="1.0" encoding="utf-8"?>
<LinearLayout xmlns:android="http://schemas.android.com/apk/res/android"
    android:orientation="vertical"
    android:layout_width="fill_parent"
    android:layout_height="fill_parent"
>
    <TextView
        android:layout_width="fill_parent"
        android:layout_height="wrap_content"
        android:text="这里是文字 1"/>
    <EditText
        android:text="这里是编辑文字 1"
        android:id="@+id/EditText01"
        android:layout_width="wrap_content"
        android:layout_height="wrap_content">
    </EditText>
    <Button
        android:text="这里是按钮 1"
        android:id="@+id/Button01"
        android:layout_width="wrap_content"
        android:layout_height="wrap_content">
    </Button>
</LinearLayout>
```

LinearLayout 的布局效果如图 4-8 所示。

图 4-8 LinearLayout 的布局效果

4.2.3 TableLayout

TableLayout 的功能是将子元素的位置分配到行或者列中。Android 中的一个 TableLayout 由许多的 TableRow 组成，每个 TableRow 都会定义一个 row。TableLayout 容器不会显示 row、cloumns 或 cell 的边框线。关于更详细的信息，可以查看这个类的 API 参考文档。下面是 TableLayout 布局代码。

TableLayout 在 Ex06_03_Layout/res/layout/table.xml 布局中的定义：

```
<?xml version="1.0" encoding="utf-8"?>
<TableLayout xmlns:android="http://schemas.android.com/apk/res/android"
    android:orientation="vertical"
    android:layout_width="fill_parent"
    android:layout_height="fill_parent"
>
    <TableRow>
      <TextView
          android:text="第一排第一个"
          android:id="@+id/TextView01">
      </TextView>
      <TextView
          android:text="第一排第二个"
          android:id="@+id/TextView02">
      </TextView>
      <TextView
          android:text="第一排第三个"
          android:id="@+id/TextView03">
      </TextView>
    </TableRow>

    <TableRow>
      <TextView
          android:text="第二排第一个"
          android:id="@+id/TextView04">
      </TextView>
      <TextView
          android:text="第二排第二个"
          android:id="@+id/TextView05">
      </TextView>
      <TextView
          android:text="第二排第三个"
          android:id="@+id/TextView06">
      </TextView>
```

```
    </TableRow>
</TableLayout>
```

在 xml 布局中，android:shrinkColumns 属性设置了表格的列是否收缩（列编号从 0 开始，下同），多列用逗号隔开（下同），如 android:shrinkColumns="0,1,2"，即表格的第 1、2、3 列的内容是收缩的，以适合屏幕，不会挤出屏幕。android:collapseColumns 属性设置了表格的列是否隐藏，android:stretchColumns 作用是设置表格的列是否拉伸。

TableLayout 的布局效果如图 4-9 所示。

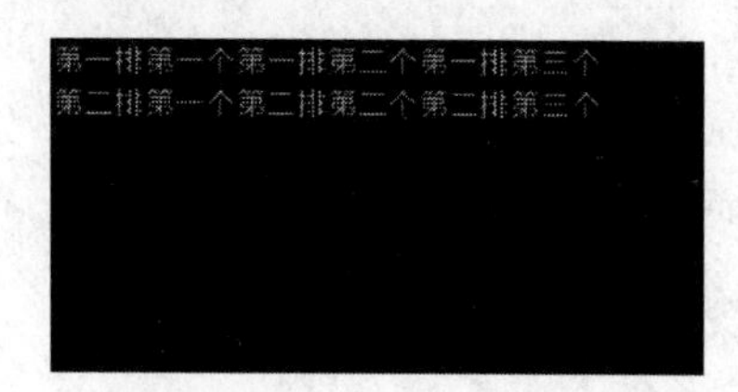

图 4-9 TableLayout 的布局效果

在默认情况下，TableLayout 是不能滚动的，超出手机屏幕的部分无法显示，在 TableLayout 的外层增加一个 ScrollView 就可以解决此问题。

4.2.4 AbsoluteLayout

AbsoluteLayout 允许以坐标的方式指定显示对象的具体位置。左上角的坐标为(0,0)，使用属性 lauout_x 和 layout_y 来指定组件的具体坐标。需要注意的是，这种布局管理器由于显示对象的位置被固定，所以在不同的设备上有可能会出现最终显示的效果不一致。下面是 AbsoluteLayout 布局代码。

AbsoluteLayout 在 Ex06_03_Layout/res/layout/absolute.xml 布局文件中的定义：

```
<?xml version="1.0" encoding="utf-8"?>
<AbsoluteLayout xmlns:android="http://schemas.android.com/apk/res/android"
    android:orientation="vertical"
    android:layout_width="fill_parent"
    android:layout_height="fill_parent">
  <Button
        android:layout_x="0dip"
        android:layout_y="100dip"
        android:text="这里是按钮 1"
        android:id="@+id/Button01"
        android:layout_width="wrap_content"
        android:layout_height="wrap_content">
  </Button>
  <TextView
        android:layout_x="10dip"
        android:layout_y="10dip"
        android:id="@+id/TextView01"
```

```
        android:text="这里是文字1"
        android:layout_width="wrap_content"
        android:layout_height="wrap_content">
    </TextView>
    <Button
        android:layout_x="150dip"
        android:layout_y="100dip"
        android:text="这里是按钮2"
        android:id="@+id/Button02"
        android:layout_width="wrap_content"
        android:layout_height="wrap_content">
    </Button>
</AbsoluteLayout>
```

AbsoluteLayout 的布局效果如图 4-10 所示。

图 4-10　AbsoluteLayout 的布局效果

4.2.5　RelativeLayout

RelatvieLayout 可以设置某一个视图相对于其他视图的位置。比如,一个按钮可以放置于另一个按钮右边,或者放在布局管理器的中央。

RelativeLayout 的属性有：

(1) android:layout_below。在某元素的下方。

(2) android:layout_top。置于指定组件之上。

(3) android:layout_toLeftOf。置于指定组件左边。

(4) android:layout_toRightOf。置于指定组件右边。

(5) android:layout_alignParentTop。与父组件上对齐。

(6) android:layout_alignParentBottom。与父组件下对齐。

(7) android:layout_alignParentLeft。与父组件左对齐。

(8) android:layout_alignLeft。与指定组件左对齐。

(9) android:layout_alignRight。与指定组件右对齐。

(10) android:layout_alignBottom。与指定组件下对齐。

(11) android:layout_alignBaseline。与指定组件基线对齐。

这些属性一部分由元素直接提供,另一部分由容器的 LayoutParams 成员(RelativeLayout 的子类)提供。RelativeLayout 参数有 Width、Height、Below、AlignTop、TopLeft、Pading 和 MarginLeft。注意，其中有些参数的值是相对于其他子元素而言的，所以才称为 RelativeLayout 布局。这些参数包括 TopLeft、AlighTop 和 Below，用来指定相对于其他元素的左、上和下的位置。下面我们来看一个 RelatvieLayout 布局。

RelatvieLayout 在 Ex06_03_Layout/res/layout/relative.xml 布局文件中的定义：

```
<?xml version="1.0" encoding="utf-8"?>
<RelativeLayout xmlns:android="http://schemas.android.com/apk/res/android"
    android:orientation="vertical"
    android:layout_width="fill_parent"
    android:layout_height="fill_parent"
>
  <TextView
        android:text="这里是文字 1"
        android:id="@+id/TextView01"
        android:layout_width="wrap_content"
        android:layout_height="wrap_content">
  </TextView>
  <Button
        android:text="这里是按钮 1"
        android:id="@+id/Button01"
        android:layout_width="wrap_content"
        android:layout_below="@+id/TextView01"
        android:layout_height="wrap_content">
  </Button>
  <TextView
        android:text="这里是文字 2"
        android:id="@+id/TextView02"
        android:layout_width="wrap_content"
        android:layout_below="@+id/Button01"
        android:layout_height="wrap_content">
  </TextView>
  <Button
        android:text="这里是按钮 2"
        android:id="@+id/Button02"
        android:layout_width="wrap_content"
        android:layout_toRightOf="@+id/Button01"
        android:layout_below="@+id/TextView01"
        android:layout_height="wrap_content">
  </Button>
</RelativeLayout>
```

RelativeLayout 的布局效果如图 4-11 所示。

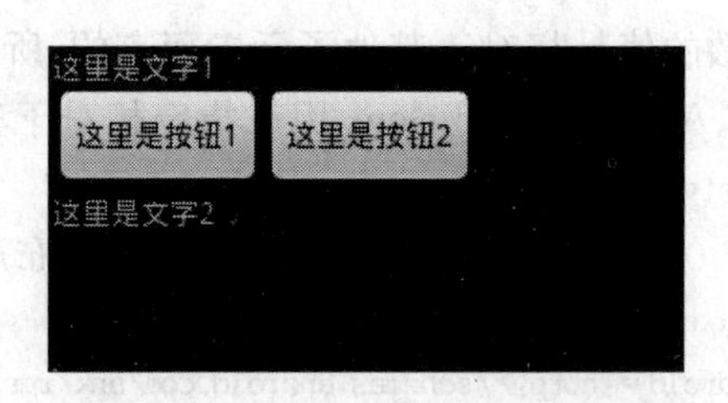

图 4-11　RelativeLayout 的布局效果

4.3　事件处理

事件是用户和 UI(图形界面)交互时所触发的操作。比如,我们按下键盘就可以触发几个事件。当键盘上的按键被按下时,触发了“按下”事件;当松开按键时,又触发了“释放”事件。在 Android 中,这些事件将被传送到事件处理器,它是一个专门接收事件对象并对其进行翻译和处理的方法。

在 Java 程序中,实现与用户交互功能的控件都需要通过事件来处理,需要指定控件所用的事件监听器。在 Android 中,同样需要设置事件监听器。另外,在 Android 下,View 同样可以响应按键和触屏两种事件。

4.3.1　事件模型

下面介绍事件源与事件监听器。

当用户与应用程序交互时,用户的操作一定是通过触发某些事件来完成的,让事件来通知程序应该执行哪些操作。这个繁杂的过程主要涉及两个对象:事件源与事件监听器。事件源指的是事件所发生的控件,各个控件在不同情况下触发的事件不尽相同,而且产生的事件对象也可能不同。监听器则是用来处理事件的对象,实现了特定的接口,根据事件的不同重写不同的事件处理方法来处理事件。

将事件源与事件监听器联系到一起,就需要为事件源注册监听,当事件发生时,系统才会自动通知事件监听器来处理相应的事件。接下来用图来说明事件处理的整个流程,如图 4-12 所示。

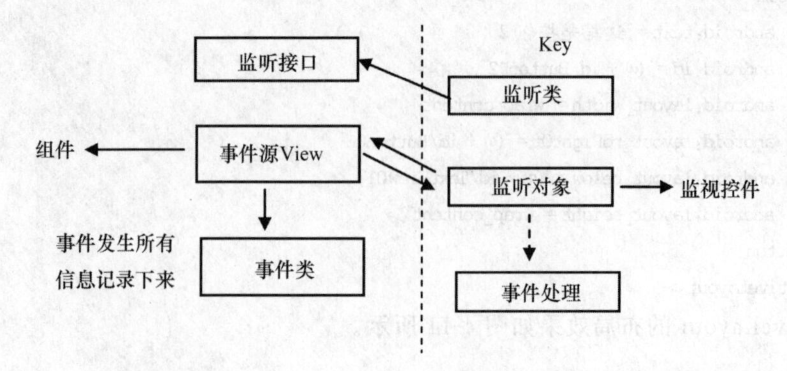

图 4-12　事件模型

4.3.2 事件监听机制

Android的事件处理机制有两种：一种是基于回调机制的，另一种是基于监听接口的。接下来分别对这两种机制进行介绍。

4.3.3 回调机制

在Android操作系统中，对事件的处理是一个非常基础而且重要的操作。许多功能的实现都需要对相关事件进行触发，然后才能达到自己的目的。比如，Android事件监听器是视图View类的接口，包含单独的回调方法。视图View类的监听器接口在视图中进行注册，当用户操作界面时触发事件，Android系统会自动调用接口中的回调方法。下面这些回调方法被包含在Android事件侦听器的接口中。

onKeyDown()：该方法是接口KeyEvent. Callback中的抽象方法，所有的View全部实现了该接口并重写了该方法，该方法用来捕捉手机键盘被按下的事件。

onKeyUp()：该方法同样是接口KeyEvent. Callback中的一个抽象方法，并且所有的View同样全部实现了该接口并重写了该方法，onKeyUp方法用来捕捉手机键盘按键释放的事件。

onTouchEvent()：该方法是在View类中的定义，并且所有的View子类全部重写了该方法，应用程序可以通过该方法处理手机屏幕的触摸事件。

onClick() ：包含在View. OnClickListener中。当用户触摸界面上的item(在触摸模式下)，或者通过浏览键或跟踪球聚焦在这个item上时，按下"确认"键或者按下跟踪球时被调用。

onLongClick() ：包含在View. OnLongClickListener中。当用户触摸并控制住界面上的item(在触摸模式下)，或者通过浏览键或跟踪球聚焦在这个item上，然后保持按下"确认"键或者按下跟踪球(一秒钟)时被调用。

onFocusChange() ：该方法包含于View. OnFocusChangeListener中。当用户使用浏览键或跟踪球浏览进入或离开界面上的item时被调用。

onKey() ：包含在View. OnKeyListener中。当用户聚焦在这个item上并按下或释放设备上的一个按键时被调用。

onTouch()：包含在View. OnTouchListener中。当用户执行的动作被当做一个触摸事件时被调用，包括按下、释放或者屏幕上任何的移动手势(在这个item的边界内)。

接下来通过简单的例子讲解onKeyDown()、onKeyUp()及onTouchEvent()方法的使用。代码如下所示。

```
publicclass Ex03_3 extends Activity {
    /** Called when the activity is first created. */
    @Override
publicvoid onCreate(Bundle savedInstanceState) {
super.onCreate(savedInstanceState);
        setContentView(R.layout.main);
        Button button = (Button)findViewById(R.id.button);
        button.setOnClickListener(new Button.OnClickListener(){
```

```
            @Override
            publicvoid onClick(View v) {
              showToast("单击按钮");
            }});
        }
        //按下按键事件
    publicboolean onKeyDown(int keyCode, KeyEvent event){
        switch(keyCode){
        case KeyEvent.KEYCODE_DPAD_UP:
            showToast("按下向上键");
            break;
        case KeyEvent.KEYCODE_DPAD_DOWN:
            showToast("按下向下键");
            break;
        case KeyEvent.KEYCODE_DPAD_LEFT:
            showToast("按下向左键");
            break;
        case KeyEvent.KEYCODE_DPAD_RIGHT:
            showToast("按下向右键");
            break;
        }
        returnfalse;
    }
        //释放按键事件
    publicboolean onKeyUp(int keyCode, KeyEvent event){
        switch(keyCode){
        case KeyEvent.KEYCODE_DPAD_UP:
            showToast("释放向上键");
            break;
        case KeyEvent.KEYCODE_DPAD_DOWN:
            showToast("释放向下键");
            break;
        case KeyEvent.KEYCODE_DPAD_LEFT:
            showToast("释放向左键");
            break;
        case KeyEvent.KEYCODE_DPAD_RIGHT:
            showToast("释放向右键");
            break;
        }
            returnfalse;
    }
    //触屏事件
    publicboolean onTouchEvent(MotionEvent event){
```

```
        int action = event.getAction();
        int posX = (int)event.getX();
        int posY = (int)event.getY();
        switch(action){
        case MotionEvent.ACTION_DOWN:
            showToast("坐标" + posX + "," + posY);
            break;
        }
            returnfalse;
    }
    publicvoid showToast(String str){
        Toast.makeText(Ex03_3.this,str,Toast.LENGTH_SHORT).show();
    }
```

当按下键盘上方向键时，界面上就会显示按下方向键，是通过 onKeyDown()方法捕捉事件的。

当松开键盘上方向键时，界面上就会显示释放方向键，是通过 onKeyUp()方法捕捉事件的。

当触摸屏幕时，通过 onTouchEvent()方法捕捉触摸屏幕的事件，根据事件获得触摸的动作，然后根据触摸动作进行处理。本例子中是把触摸屏幕的光标的 x、y 坐标显示到屏幕上。

第5章 高级UI设计

本章介绍 Android 平台应用程序高级 UI 的设计方法，包括 Menu，ListView，Spinner，Gallery，Toast，AlertDialog 等。

5.1 Menu

Android 手机专门用一个按键“Menu”来显示菜单，它的功能是在屏幕底部弹出一个菜单，这个菜单我们就叫它选项菜单 OptionsMenu。一般情况下，选项菜单最多显示 2 排，每排 3 个菜单项。这些菜单项有文字和图标，故也被称做 Icon Menus。如果可选择项多于 6 项，第六项以后的选择项会被隐藏，第六项被“More”取代，单击“More”才出现第六项及以后的菜单项，这些菜单项也被称做 Expanded Menus。所以，如果应用程序设置了菜单，我们便可以通过该按键来操作应用程序的菜单选项。

要实现菜单功能，首先需要通过方法 onCreateOptionsMenu 来创建菜单，然后需要对能够触发的事件进行监听，这样才能够在事件监听 onOptionsItemSelected 中根据不同的菜单选项来执行不同的任务。

我们通过一个例子来讲解如何实现 Menu 菜单，通过 Menu 菜单切换图片。

首先，我们需要定义菜单按钮的 id，以便于以后修改。

```
/** 定义菜单 id */
privatestaticfinalintM_CHANGE_FIRST = Menu.FIRST; //切换第一张图
privatestaticfinalintM_CHANGE_SECOND = Menu.FIRST + 1; //切换第二张图
privatestaticfinalintM_HELP = Menu.FIRST + 2;//帮助
```

然后，重写 onCreateOptionsMenu()方法来定制我们的菜单，在此方法中，我们通过 Menu 对象创建菜单。实现代码如下所示。

```
/** 创建 Menu 菜单，重写 onCreateOptionsMenu 方法 */
  @Override
  publicboolean onCreateOptionsMenu(Menu menu) {
        //创建 Menu 群组 id
        int idGroup1 = 0;
```

```
        //创建 Menu 顺序 id
        int orderMenuItem1 = Menu.NONE;
        int orderMenuItem2 = Menu.NONE + 1;
        int orderMenuItem3 = Menu.NONE + 2;
        menu.add(idGroup1,M_CHANGE_FIRST,orderMenuItem1,"切换第一张图");
        menu.add(idGroup1,M_CHANGE_SECOND,orderMenuItem2,"切换第二张图");
        menu.add(idGroup1,M_HELP,orderMenuItem3,"帮助");
        returnsuper.onCreateOptionsMenu(menu);
    }
```

menu.add()方法的第一个参数表示对菜单进行分组;第二个参数是菜单的id,是菜单的唯一标识;第三个参数表示菜单的显示顺序;第四个参数是文本,表示要显示的菜单文字。

当菜单显示出来后,用户单击菜单中的某一项,这时菜单需要响应这个单击事件。这也很简单,可以通过重载onOptionsItemSelected()方法来实现,如下面的代码所示:

```
    /** 选择 Menu 菜单,重写 onOptionsItemSelected 方法 */

    @Override
    publicboolean onOptionsItemSelected(MenuItem item) {
        int id = item.getItemId();//获得 Menu 菜单的 id
        switch(id){
        caseM_CHANGE_FIRST:
imageView.setImageDrawable(getResources().getDrawable(R.drawable.a));
        break;
        caseM_CHANGE_SECOND:
        imageView.setImageDrawable(getResources().getDrawable(R.drawable.b));
            break;
        caseM_HELP:
            Intent intent = new Intent(MainActivity.this,HelpActivity.class);
            startActivity(intent);
            break;
        }
        returnsuper.onOptionsItemSelected(item);
    }
```

程序的效果运行如图5-1所示。

图 5-1　Menu 菜单的效果

上下文菜单是注册到某个 View 对象上的。如果一个 View 对象注册了上下文菜单，用户可以通过长按该 View 对象调出上下文菜单。我们来看一个关于 ContextMenu 的例子，这个例子是关于如何查看和删除信息的。

我们先创建一个 ContextMenu，它需要重写 onCreateContextMenu 方法，在 onCreateContextMenu()方法里调用 Menu 的 add 方法添加菜单项(MenuItem)，代码如下所示。

```
/**
    * 创建长按菜单(上下文菜单)
    * @param ContextMenu menu
    * @param View v
    */
  @Override
  publicvoid onCreateContextMenu(ContextMenu menu, View v,
                                                    ContextMenuInfo menuInfo) {
        menu.add(0,M_UPDATE,0,"修改");//添加修改菜单项
        menu.add(0,M_DEL,0,"删除");//添加删除菜单项
        menu.add(0,M_SELECT,0,"查看");//添加查看菜单项
        super.onCreateContextMenu(menu, v, menuInfo);
  }
```

接下来处理菜单的单击事件，需要先重写 onContextItemSelected 方法，然后通过参数 item 获得菜单 id 并进行判断，实现代码如下所示。

```
/**
   * 长按菜单(上下文菜单)响应事件
```

```
    * @param MenuItem item
    */
  @Override
  publicboolean onContextItemSelected(MenuItem item) {
          //获得当前被选择的菜单项信息
      AdapterContextMenuInfo info =
                      (AdapterContextMenuInfo)item.getMenuInfo();
      long itemID = info.id;
      Intent intent; //Intent 对象
      switch(item.getItemId())
      {
      caseM_UPDATE:
                  //创建 Intent 对象
        intent = new Intent(Diary.this,EditDiary.class);
        intent.putExtra("option", 2); //传递数据
        intent.putExtra("itemID", itemID);//传递数据
        startActivity(intent);//启动一个新的 Activity
        dao.close();//关闭数据库对象
        Diary.this.finish();//关闭当前的 Activity
        break;
      caseM_DEL:
        dao.delete(itemID);//数据库对象调用删除方法
        readerListView();//调用读取列表方法
        break;
      caseM_SELECT:
        intent = new Intent(Diary.this,EditDiary.class);
        intent.putExtra("option", 1);
        intent.putExtra("itemID", itemID);
        startActivity(intent);
        dao.close();
        Diary.this.finish();
        break;
      }
      returnsuper.onContextItemSelected(item);
  }
```

完成上面的两步操作之后,还需要注册菜单,注册菜单的方法是 registerForContext-Menu。如果我们不注册上下文菜单,将导致我们的上下文菜单不能显示。

```
    /** 注册长按菜单 */
    registerForContextMenu(getListView());
```

程序的运行效果如图 5-2 所示。

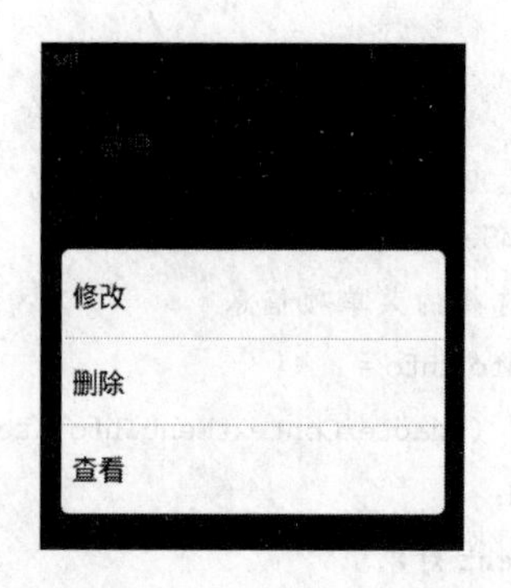

图 5-2 ContextMenu 菜单的效果

5.2 ListView

在 Android 开发中，ListView 是比较常用的组件，它以列表的形式展示具体的内容，并且能够根据数据的长度自适应显示。

列表的显示需要 3 个元素：

(1) ListVeiw。用来展示列表的 View。

(2) 适配器。用来把数据映射到 ListView 上的中介。

(3) 数据。将被映射的字符串、图片、基本组件等资源。

根据列表的适配器类型，列表可分为 3 种：ArrayAdapter、SimpleAdapter 和 SimpleCursorAdapter，其中以 ArrayAdapter 最为简单，只能显示一行字。SimpleAdapter 有最好的扩充性，可以自定义出各种效果。SimpleCursorAdapter 可以认为是 SimpleAdapter 对数据库的简单结合，可以方便地把数据库的内容以列表的形式展示出来。

接下来，我们来看 3 个关于 listView 的例子。

第一个例子是用 listView 来显示普通的列表，在这个例子中，我们首先需要在 Activity 类中创建一个 ListView 对象，然后定义要添加到 listView 中的数据，具体如下面的代码所示。

```
 //创建 ListView 对象
 private ListView listView;
//添加数据
 private List<String> getData(){
     List<String> data = new ArrayList<String>();
     data.add("唐僧");
     data.add("悟空");
     data.add("沙僧");
     data.add("八戒");
     return data;
 }
```

定义好数据后，创建一个 ArrayAdapter，将数据与 listView 里显示的布局结合起来，代码如下所示：

```
publicvoid onCreate(Bundle savedInstanceState){
  super.onCreate(savedInstanceState);
  listView = new ListView(this);  //获得 listView 组件
  //设置列表的适配器
  listView.setAdapter(new ArrayAdapter<String>(this, android.R.layout.simple_expand-
able_list_item_1,getData()));
}
```

最后，我们将做好的 listView 通过 setContentView()方法显示到界面上，代码如下：

```
setContentView(listView);  //将 listView 显示到当前页面
```

我们来看一下效果，如图 5-3 所示。

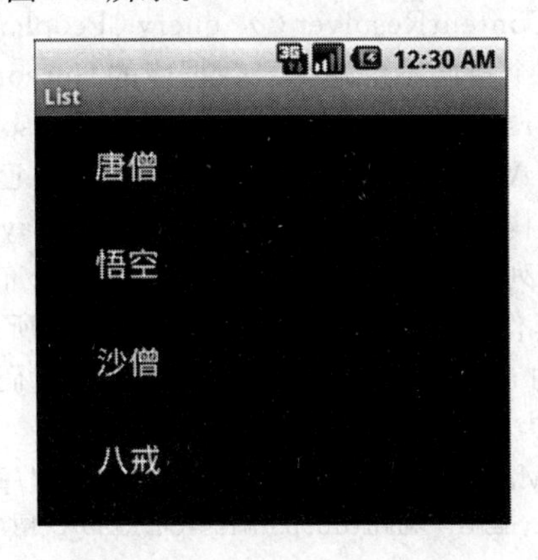

图 5-3　listView 列表的效果

上面代码使用了 ArrayAdapter(Context context, int textViewResourceId, List<T> objects)来装配数据，要装配这些数据就需要一个连接 ListView 视图对象和数组数据的适配器来做两者的适配工作。ArrayAdapter 的构造需要 3 个参数，第一个参数是上下文，第二个参数是指定列表项的模板，也就是一个 xml 布局文件的资源 ID，第三个参数是列表项中显示数据，然后使用 setAdapter()来完成适配阶段的最后工作。

注意：上下文在这个例子里指的是当前 Activity 的对象实例(this)；该代码中布局资源 ID：android.R.layout.simple_list_item_1 是系统定义好的布局文件，它只显示一行文字，它可以在 Android SDK 安装目录下 platforms\android-2.1\data\res\layout 目录中找到。

第二个例子是用 listView 来显示联系人的列表信息。在这个例子中，我们将采用 SimpleCursorAdapter。先在通信录中添加一个联系人作为数据库的数据，然后获得一个指向数据库的 Cursor 游标对象并且定义一个布局文件(当然也可以使用系统自带的)，具体代码如下所示。

```
public class MyListView2 extends Activity {
//定义 ListView 对象
private ListView listView;
@Override
public void onCreate(Bundle savedInstanceState){
    super.onCreate(savedInstanceState);
    listView = new ListView(this);
```

```
        //获取系统联系人信息
        Cursor cursor =  getContentResolver().query(People.CONTENT_URI, null, ull, null, null);
        startManagingCursor(cursor);
    ListAdapter listAdapter = new SimpleCursorAdapter(this,
     android.R.layout.simple_expandable_list_item_1,
    cursor,new String[]{People.NAME},new int[]{android.R.id.text1});
        //设置列表的适配器
    listView.setAdapter(listAdapter);
        setContentView(listView);
      }
    }
```

C ursor cursor = getContentResolver ( ). query (People. CONTENT_ URI, null, null, null, null);先获得一个指向系统通信录数据库的 Cursor 对象,从而获得数据来源。

通过 startManagingCursor(cursor),我们将获得的 Cursor 对象交由 Activity 管理,这样 Cursor 的生命周期和 Activity 便能够自动同步,省去自己手动管理 Cursor 的工作。

SimpleCursorAdapter 构造函数的前面 3 个参数与 ArrayAdapter 是一样的,最后两个参数:一个是包含数据库列的 String 型数组,一个是 xml 布局文件中组件标签的 android:id 属性值的数组。其作用是自动地将 String 型数组所表示的每一列数据映射到 xml 布局文件中组件标签的 android:id 属性值。在上面的代码中,是将 NAME 列的数据依次映射到布局文件的 id 为 text1 的组件上。

注意,需要在 AndroidManifest. xml 中加入权限,否则程序运行会报错。

```
<uses-permission android:name="android.permission.READ_CONTACTS">
</uses-permission>
```

程序运行效果如图 5-4 所示。

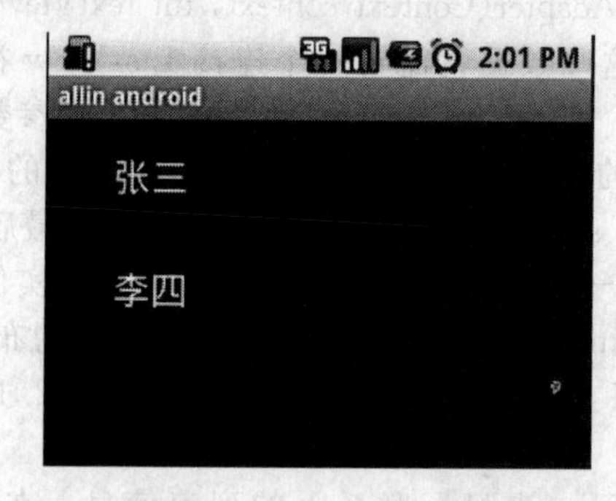

图 5-4　显示联系人列表

第三个例子是关于 simpleAdapter 的使用。simpleAdapter 的扩展性最好,可以定义各种各样的布局,可以放上 ImageView,还可以放上 Button 和 CheckBox 等。

下面的程序实现了一个带有图片的类表。首先定义好一个用来显示每一列内容的布局文件 vlist. xml:

```
<LinearLayout xmlns:android="http://schemas.android.com/apk/res/android"
    android:orientation="horizontal" android:layout_width="fill_parent"
    android:layout_height="fill_parent">
<ImageView android:id="@+id/img"
      android:layout_width="wrap_content"
```

```
        android:layout_height = "wrap_content"
        android:layout_margin = "5px"/>
<LinearLayout android:orientation = "vertical"
        android:layout_width = "wrap_content"
        android:layout_height = "wrap_content">
  <TextView android:id = "@ + id/title"
            android:layout_width = "wrap_content"
            android:layout_height = "wrap_content"
            android:textColor = "#FFFFFFFF"
            android:textSize = "22px" />
  <TextView android:id = "@ + id/info"
            android:layout_width = "wrap_content"
            android:layout_height = "wrap_content"
            android:textColor = "#FFFFFFFF"
            android:textSize = "13px" />
  </LinearLayout>
  </LinearLayout>
```

Simple 类继承 ListActivity，ListActivity 是 Activity 的子类。在 Simple 类里将数据加载到 list 集合中，然后定义 simpleAdapter 适配器将 list 集合中的数据和布局文件结合起来。

```
publicclass Simple extends ListActivity {
publicvoid onCreate(Bundle savedInstanceState) {
super.onCreate(savedInstanceState);
        SimpleAdapter adapter = new SimpleAdapter(this,getData(),R.layout.vlist,
new String[]{"title","info","img"},
newint[]{R.id.title,R.id.info,R.id.img});
        setListAdapter(adapter);
    }
private List<Map<String, Object>> getData() {
        List<Map<String, Object>> list = new ArrayList<Map<String, Object>>();
        Map<String, Object> map = new HashMap<String, Object>();
        map.put("title", "G1");
        map.put("info", "google 1");
        map.put("img", R.drawable.tool0);
        list.add(map);
        map = new HashMap<String, Object>();
        map.put("title", "G2");
        map.put("info", "google 2");
        map.put("img", R.drawable.tool1);
        list.add(map);
        map = new HashMap<String, Object>();
        map.put("title", "G3");
        map.put("info", "google 3");
        map.put("img", R.drawable.tool2);
        list.add(map);
```

```
return list;
    }
}
```

程序的运行效果如图 5-5 所示。

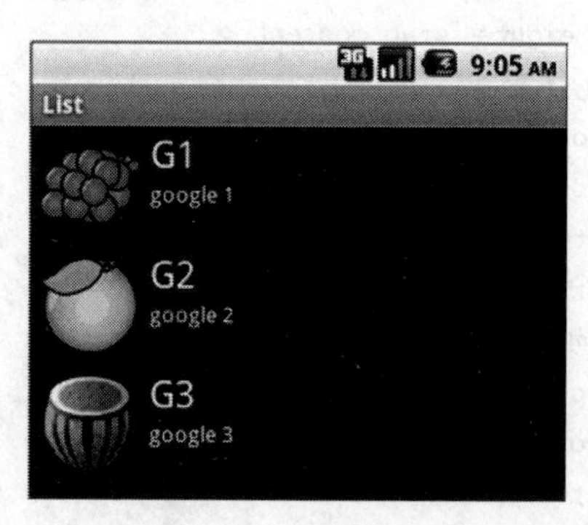

图 5-5　水果列表

使用 simpleAdapter 的数据一般都是由 HashMap 构成的 List，List 的每一节对应 ListView 的每一行。HashMap 的每个键值数据映射到布局文件中对应 id 的组件上。因为系统没有对应的布局文件可用，我们可以自己定义一个布局文件（如上面例子中的 vlist. xml）。SimpleAdapter 构造方法里的第一个参数是上下文，当前 Activity 的对象实例（this）；第二个参数是数据；第三个参数是模板的资源 ID；第四个参数是组件对应的资源；第五个参数是 xml 布局文件中组件的 id。布局文件的各组件分别映射到 HashMap 的各元素上，完成适配。

5.3　Spinner

当我们在网站上注册账号时，网站可能会让我们提供性别、生日、城市等信息。网站开发人员为了方便用户，不让用户填写这些信息，而是提供一个下拉列表将所有的可选项列出来，让用户选择。Android 给我们提供了一个 Spinner 控件，这个控件主要就是一个列表，Spinner 位于 android. widget 包下，每次只显示用户选中的元素，当用户再次单击时，会弹出选择列表供用户选择，而选择列表中的元素同样来自适配器。

我们来看一个选择居住地区的例子。在例子中，我们先在 Layout 下的 xml 中添加一个 Spinner 组件，代码如下：

```
<? xml version = "1.0" encoding = "utf-8" ? >
<LinearLayout xmlns:android = "http://schemas.android.com/apk/res/android"
  android:orientation = "vertical" android:layout_width = "fill_parent"
  android:layout_height = "fill_parent">
  <TextViewandroid:id = "@ + id/tv" android:layout_width = "fill_parent"
    android:layout_height = "wrap_content" android:text = "@string/hello" />
  <Spinnerandroid:id = "@ + id/sp" android:layout_width = "fill_parent"
    android:layout_height = "wrap_content" />
</LinearLayout>
```

然后，我们需要在 Activity 类中获得 Spinner 组件：

```
        spinner = (Spinner)findViewById(R.id.sp);
        textView = (TextView)findViewById(R.id.tv);
```

接下来，我们需要创建一个适配器，此适配器需要是 ArrayAdapter，代码如下：

```
/*
     * new ArrayAdapter 对象并将 allCity 传入
     **/
   adapter = new ArrayAdapter<String>(this, android.R.layout.simple_spinner_item,allCity);
```

定义好适配器后，将适配器设置给 Spinner 组件，并为其注册监听事件以弹出提示信息，代码如下：

```
  //将 adapter 添加到 spinner 中
        spinner.setAdapter(adapter);

        spinner.setOnItemSelectedListener(new OnItemSelectedListener(){
          @Override
          public void onItemSelected(AdapterView<? > arg0, View arg1,
              int arg2, long arg3) {
            textView.setText("你选择的是" + city[arg2]);
          }
          @Override
          public void onNothingSelected(AdapterView<? > arg0) {
            // TODO Auto-generated method stub
          }});
    }
```

程序运行效果如图 5-6 所示。

图 5-6　Spinner 效果图

5.4 Gallery

还记得 iPhone 中用手指拖动图片的效果吗？iPhone 曾经凭借这个效果吸引了不少眼球。要在 Android 平台上实现这样的效果，需要一个容器来存放 Gallery 显示的图片，这里使用一个继承自 BaseAdapter 类的派生类来显示这些图片。我们需要监听其事件 setOnItemClickListener，从而确定用户当前选中的是哪一张图片。首先，需要将所有要显示的图片的索引存放在一个 int 型数组中，然后通过 setImageResource 方法来设置 ImageView 要显示的图片资源，最后将每张图片的 ImageView 显示在屏幕上。下面的例子实现了这一过程，当单击某张图片的时候，捕捉并处理该事件。

先看看存放这些图片的容器 ImageAdapter 的实现：

```
publicclass ImageAdapter extends BaseAdapter
    {
        /*声明变量*/
int mGalleryItemBackground;
private Context mContext;
        /*ImageAdapter 的构造器*/
public ImageAdapter(Context c)
        {
          mContext = c;
        }
        /* 覆盖的方法 getCount,返回图片数目 */
publicint getCount()
        {
return myImageIds.length;
        }
        /* 覆盖的方法 getItemId,返回图像的数组 id */
public Object getItem(int position)
        {
return position;
        }
publiclong getItemId(int position)
        {
return position;
        }
        /* 覆盖的方法 getView,返回一 View 对象 */
public View getView(int position, View convertView, ViewGroup parent)
        {
          /*产生 ImageView 对象*/
          ImageView i = new ImageView(mContext);
          /*设置图片给 imageView 对象*/
```

```
        i.setImageResource(myImageIds[position]);
        /*重新设置图片的宽和高*/
        i.setScaleType(ImageView.ScaleType.FIT_XY);
        /*重新设置 Layout 的宽和高*/
        i.setLayoutParams(new Gallery.LayoutParams(240, 240));
        /*设置 Gallery 的背景图*/
        i.setBackgroundResource(mGalleryItemBackground);
        /*返回 imageView 对象*/
return i;
      }
      /*建构 Integer array,并取得预加载 Drawable 的图片 id*/
private Integer[] myImageIds =
      {
          R.drawable.hou,R.drawable.zhu,R.drawable.tang
      };
}
```

然后通过 setAdapter 方法把资源文件添加到 Gallery 中显示，并设置其事件处理，代码如下：

```
publicclass Ex05_8 extends Activity {
  private Gallery ga;
    /** Called when the activity is first created. */
    @Override
publicvoid onCreate(Bundle savedInstanceState) {
super.onCreate(savedInstanceState);
        setContentView(R.layout.main);
        ga = (Gallery)findViewById(R.id.ga);
        ga.setAdapter(new ImageAdapter(this));

        ga.setOnItemClickListener(new OnItemClickListener(){
            @Override
            publicvoid onItemClick(AdapterView<?> arg0, View arg1, int id,
                    long arg3) {
              Toast.makeText(Ex05_8.this,"now:"+ id,Toast.LENGTH_SHORT).show();
            }});
}
...
```

程序的运行效果如图 5-7 所示。

图 5-7　图片浏览

5.5　Toast

Toast 是 Android 提供的“提示信息”类，Toast 类的使用非常简单，而且用途很多。比如在要求用户输入年龄的时候，不小心输入了字母，这个时候 Toast 可以提示用户“只能输入数字”。下面我们通过一个示例来演示 Toast 的用法。

```
public class Ex05_1 extends Activity {
      private Button button;
   /** Called when the activity is first created. */
   @Override
   public void onCreate(Bundle savedInstanceState)
{
      super.onCreate(savedInstanceState);
   setContentView(R.layout.main);
   button = (Button)findViewById(R.id.button);
   button.setOnClickListener(new Button.OnClickListener()
{
      @Override
      public void onClick(View v)
{
        Toast.makeText(Ex05_1.this,"你单击了一下按钮",Toast.LENGTH_SHORT).show();
      }
});
```

```
    }
  }
```

当单击了按钮之后,屏幕会弹出 Toast 信息,Toast 是没有焦点的,而且 Toast 显示的时间有限,过了一定的时间就会自动消失。在上面的示例代码中,我们用 Toast. LENGTH_SHORT 对时间作了限制。

Toast 的 makeText 方法里有 3 个参数,分别对应的是:Context 上下文、Toast 显示的内容,以及显示时间。如果需要显示的时间很短,可以用 Toast. LENGTH_SHORT;如果需要显示时间很长,则可以用 Toast. LENGTH_LONG。

程序的运行效果如图 5-8 所示。

图 5-8　Toast 的效果图

5.6　AlertDialog

我们在与用户交互的过程中,经常会用到“确认”、“警告”这样的 Dialog 形式的提示窗口。本节介绍的是 AlertDialog 提示窗口,它也是 Android 中创建对话框最常用的方法。

在下面的例子中,当我们单击其中一个按钮后,会弹出一个 AlertDialog 对话窗口。在这个例子的实现代码中,通过 setPositiveButton 和 setNegativeButton 方法加入了两个按钮,每个按钮都有它自己的监听事件。如果对话框设置了按钮,那么需要对其设置事件监听 OnClickListener。这个例子构造了一个具有标题(setTitle)、提示信息 Message(setMessage)和两个按钮的对话窗口。另外还可以通过 setIcon()方法设置对话框里的图标,比如 setIcon(R. drawable. myIcon)。

看看本例中 AlertDialog 的实现代码,如下所示:

```
publicclass Ex05_6_alert extends Activity
{
private Button button;
  /** Called when the activity is first created. */
  @Override
publicvoid onCreate(Bundle savedInstanceState)
  {
super.onCreate(savedInstanceState);
    setContentView(R.layout.alert);
    button = (Button) findViewById(R.id.bt);
    button.setOnClickListener(new Button.OnClickListener(){
publicvoid onClick(View v) {
new AlertDialog.Builder(Ex05_6_alert.this).setTitle("about")
```

```
            .setMessage("你真有勇气!")
            .setPositiveButton("确定", new DialogInterface.OnClickListener(){
    publicvoid onClick(DialogInterface di,int i){}
            })
            .setNegativeButton("百度首页", new DialogInterface.OnClickListener(){
    publicvoid onClick(DialogInterface di,int i){
                Uri uri = Uri.parse("http://www.baidu.com");
                Intent intent = new Intent(Intent.ACTION_VIEW,uri);
                startActivity(intent);
            }
            }).show();
        }});
    }
}
```

由于 AlertDialog 类的不能直接使用 new 关键字来创建 AlertDialog 类的对象实例。为了能创建 AlertDialog 对象，需要调用 AlertDialog 的内部类 Builder 类。创建完 AlertDialog 对象后需要通过 Builder 类的 show()方法显示对话框。

程序的运行效果如图 5-9 所示。

图 5-9　图片浏览

图 5-9 对话框接下来看看另一种对话框 ProgressDialog，与 AlertDialog 不同的是，ProgressDialog 显示的是一种"加载中"的效果。

需要注意的是，Android 的 ProgressDialog 必须要在后台程序运行完毕前以 dismiss()方法来关闭取得焦点(focus)的对话框，否则程序会陷入无法终止的死循环中。在线程里不能有任何更改当前 Activity 或父视图的任何状态，文字输出等事件，因为线程里的当前 Activity 与视图并不属于父视图，两者之间也没有关联。在接下来的例子中，如果你想在线程里改变图片的 Alpha 值，是没有效果的。在该示例中，我们通过一个线程来模拟后台程序的运行，在线程完毕时，关闭这个加载中的动画对话框。

ProgressDialog 的实现代码如下所示：

```
public class Ex05_6_progress extends Activity
{
   private Button button;
   //创建 ProgressDialog 对象
   private ProgressDialog myDialog = null;
   /** Called when the activity is first created. */
   @Override
   public void onCreate(Bundle savedInstanceState)
   {
super.onCreate(savedInstanceState);
       setContentView(R.layout.progress);
       button = (Button) findViewById(R.id.bt);
button.setOnClickListener(myShowProgressBar);
   }
   //Button 监听事件
   Button.OnClickListener myShowProgressBar =
   new Button.OnClickListener()
   {
       public void onClick(View arg0)
       {
           final CharSequence strDialogTitle =
                   getString(R.string.str_dialog_title);
           final CharSequence strDialogBody =
                   getString(R.string.str_dialog_body);
   // 显示 Progress 对话框
   myDialog = ProgressDialog.show
                   (
                       Ex05_6_progress.this,
                       strDialogTitle,
                       strDialogBody,
                       true
                   );
   new Thread()
   {
       public void run()
       {
   try
   {
       sleep(1000);
   }
   catch (Exception e)
   {
```

```
        e.printStackTrace();
    }
    finally
    {
        // 卸载所创建的 myDialog 对象
        myDialog.dismiss();
    }
        }
    }.start(); /* 开始运行线程 */
        }
    };
}
```

程序的运行效果如图 5-10 所示。

图 5-10　进度条对话

第6章 GPA计算能手项目案例

GPA(Grade Point Average)计算能手是一款用来计算学分和绩点的软件,它属于应用类软件。本章以GPA计算能手项目为例介绍应用类软件UI的一般设计方法,以及本软件所涉及的数据输入、数值计算、数据处理、数据显示等内容。

6.1 预备知识

- Activity。Activity的创建和生命周期的使用,通过Intent跳转Activity,并完成内容的传递。
- 布局管理器的使用。用到五大布局中的3种,分别是FrameLayout、LinearLayout和RelativeLayout。
- 几种基本组件的使用。包括高级组件中的ListView、Toast、AlertDialog。
- SQLite数据库。SQLite数据库是一个开源的关系型数据库,与普通的关系型数据库一样,也具有ACID即改查增删等特性,它可以用来存储大量的数据,能够很容易地对数据执行查询、更改、删除等操作。
- 适配器Adapter的使用。得到数据和视图之后通过合适的适配器,将两者结合起来。

另外,除了数据库,还有另外一种存储数据的方法——SharedPreferences,用于保存系统的配置信息。比如,如果我们想让用户在应用下次启动时自动登录,就需要它将输入的用户名和密码保存起来。

6.2 需求分析

近些年,随着中国在教育领域与国外交流的日益增多,国外高中和大学的优质教育资源和人性化教育理念吸引了越来越多的中国学生,他们开始倾向于选择进入国外高中和大学进行学习和深造。

每一个有意出国留学的学生都面临着一个不可回避的问题,即国内取得成绩与国外认可成绩之间的换算问题。这就是留学生中经常提及的GPA换算。GPA成绩是申请国外大学的最重要的一个材料,因此GPA的换算已成为所有出国学生最为关心一个问题。

由于目前国内不论是高中还是大学，成绩计算方法种类繁多，同样一个成绩根据不同学校的计算后，结果差距明显，这就给学生准确自身定位和申请国外理想大学造成了严重障碍。

6.3 功能分析

本款基于 Android 平台的 GPA 计算能手软件为此问题提供了专业的解决方案。GPA 计算能手软件提供了国外高校认可的多种 GPA 算法，帮助用户把国内高校的成绩准确换算为国外教育机构认可的 GPA 成绩。使用户对自己的成绩有一个准确的认识，解决在出国时成绩方面的困惑。

6.4 设　　计

GPA 计算能手软件的界面设计与功能描述如表 6-1 所示。

表 6-1　GPA 计算能手软件的界面设计与功能描述

界面	功能简述
主界面	1. 信息显示 (1)多种 GPA 算法结果显示 (2)已获得学分 (3)当前学期，学年时间显示 2. 功能按钮 (1)输入成绩 (2)查询成绩 (3)未来成绩计算 (4)导出数据 (5)更多
输入信息	1. 输入信息 (1)课程名称 (2)学期 (3)类型 (4)学分 (5)成绩 (6)备注 2. 功能按钮 (1)返回按钮 (2)存储返回按钮 (3)存储继续输入按钮

续表

界面	功能简述
查询数据	1. 每门信息显示 (1)名称 (2)学期 (3)类型 (4)成绩 (5)学分 2. 功能按钮 (1)返回按钮 (2)添加新课程按钮 (3)删除按钮 (4)筛选按钮(学期 / 类型) (5)长按修改
未来计算	1. 两种方式信息提示 2. 输入数据 (1)待获得学分 (2)需要达到的目标成绩
更多	1. 关于 2. FAQ 常见问题 3. 云端上传功能

6.4.1　UI 设计

(1) Logo 界面如图 6-1 所示。

(2) 主界面如图 6-2 所示。

图 6-1　Logo 界面

图 6-2　主界面

（3）输入数据的界面如图 6-3 所示。

（4）查询界面如图 6-4 所示。

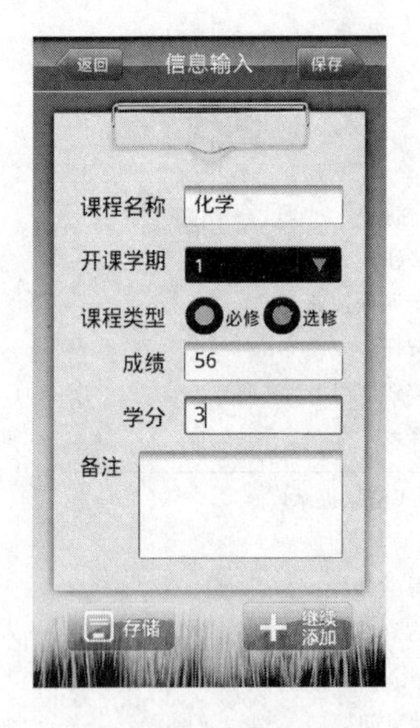

图 6-3　输入数据界面

图 6-4　查询界面

（5）未来计算界面如图 6-5 所示。

（6）其他界面如图 6-6 所示。

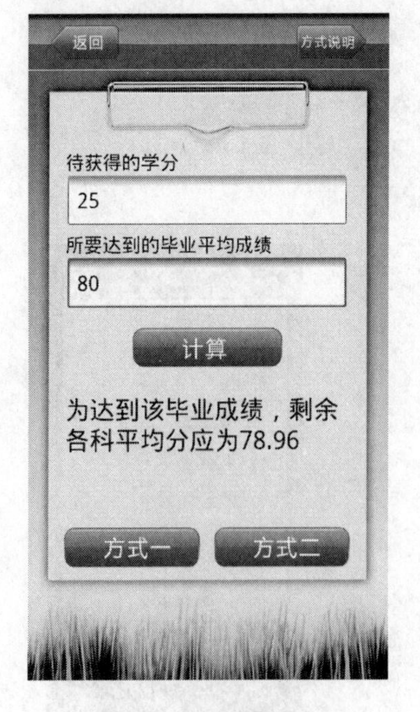

图 6-5　未来计算界面

图 6-6　其他界面

（7）“关于”界面如图 6-7 所示。

（8）“FAQ”界面如图 6-8 所示。

图 6-7　“关于”界面

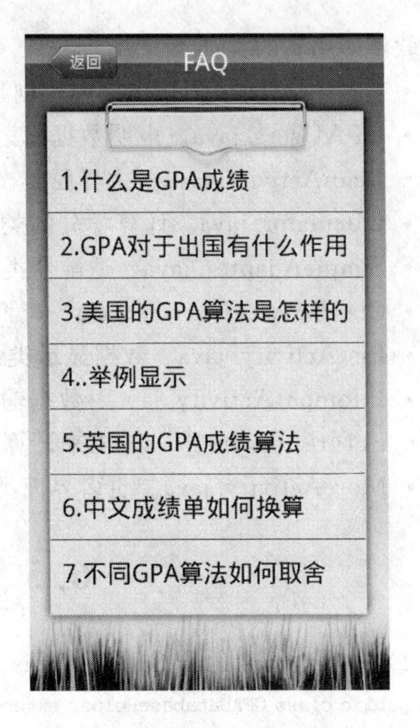

图 6-8　“FAQ”界面

（9）FAQ 问题解答界面如图 6-9 所示。

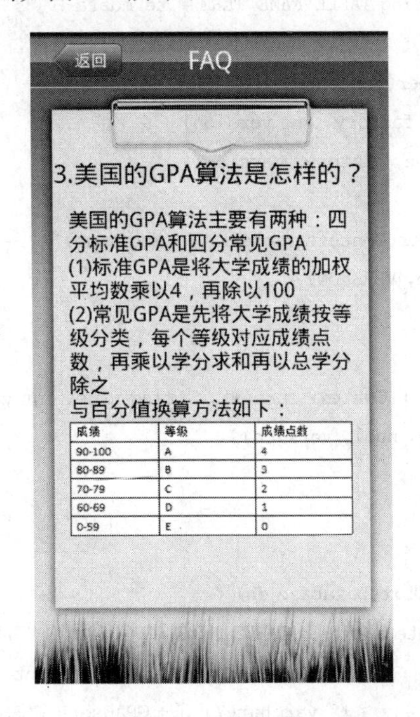

图 6-9　FAQ 解答界面

6.4.2 类设计

软件中用到的类如下：

- GPADatabaseHelper. java。创建数据库的类。
- GPAUsers. java。声明数据库语句中字段的类。
- MainActivity. java。主界面类。
- Calculator. java。计算学分的类。
- SimpleAdapter. java。适配器类。
- ClassbrowerActivity. java。查询数据的界面的类。
- ListActivity. java。数据显示类。
- InfoinputActivity. java。数据输入类。
- FutureActivity. java。未来计算类。
- MoreActivity. java。更多界面类。

6.5 编程实现

(1) 创建数据库的类 GPADatabaseHelper：

```
public class GPADatabaseHelper extends SQLiteOpenHelper {
public static final int VERSION = 1;
public static final String TABLE_NAME_GPA = "GPAdata";
public static final String TABLE_NAME_TERM = "termdata";

public GPADatabaseHelper(Context context, String name,
        CursorFactory factory, int version) {
    super(context, name, factory, version);
}
public GPADatabaseHelper(Context context, String name) {
    this(context, name, VERSION);
}

public GPADatabaseHelper(Context context, String name, int version) {
    this(context, name, null, version);
}

@Override
public void onCreate(SQLiteDatabase db) {
      db.execSQL("create table " + TABLE_NAME_GPA + " ( " + GPAUsers._ID + "
                              Integer primary key autoincrement, " + GPAUsers.USERNAME
                                + " varchar(7) ," + GPAUsers.CLASSNAME + " vachar(10),"
                                              + GPAUsers.CLASSTAG + " integer ,"
```

```
                                + GPAUsers.TERM + " integer," + GPAUsers.SCORE
                                    + " float ," + GPAUsers.CREDIT + " float ,"
                                        + GPAUsers.OTHER + " varchar(20),"
                                            + GPAUsers.TAG + " integer);");
        db.execSQL("create table " + TABLE_NAME_USER + " (" + GPAUsers._ID
                + " Integer primary key autoincrement, " + PAUsers.USERNAME
                + " varchar(7) ," + GPAUsers.TERM + " integer);");

    }

    @Override
    public void onUpgrade(SQLiteDatabase db, int oldVersion, int newVersion) {
    }

    }
```

(2) GPAUsers.java 是用来声明数据库语句中字段的一个类，它实现了 BaseColumns 接口：

```
    public class GPAUsers implements BaseColumns {
        public static final String USERNAME = "username";
        public static final String CLASSNAME = "classname";
        public static final String CLASSTAG = "classtag";
        public static final String TERM = "term";
        public static final String SCORE = "score";
        public static final String CREDIT = "credit";
        public static final String OTHER = "other";
        public static final String TAG = "tag";
    }
```

(3) 应用加载时的界面和进入应用之后的主界面都由 MainActivity.java 中的代码完成，通过为 Activity 设置不同的布局来实现，如何改变布局呢？通过一个变量 resumetag，初始值是 0，在 onResume()方法中判断如果是 0 则显示 Logo 界面：

```
public class MainActivity extends Activity {
    private TextView xueqi = null;
    private TextView xuenian = null;
    private TextView gpa1 = null;
    private TextView gpa2 = null;
    private TextView pingjun = null;
    private TextView yihuo = null;
    private ImageButton exit = null;
    private Button shurushuju = null;
    private ImageButton chaxun = null;
    private ImageButton weilai = null;
    private ImageButton daochu = null;
```

```
    private ImageButton shezhi = null;

    int resumetag = 0;// 判断显示哪个界面,0-显示 Logo 界面,1-显示主界面

    GPADatabaseHelper dbhelp = new GPADatabaseHelper(MainActivity.this,
            GPADatabaseHelper.TABLE_NAME_TERM,
                        GPADatabaseHelper.VERSION);// 数据库

    @Override
    public void onCreate(Bundle savedInstanceState) {
        super.onCreate(savedInstanceState);
        requestWindowFeature(Window.FEATURE_NO_TITLE);
        getWindow().setFlags(WindowManager.LayoutParams.FLAG_FULLSCREEN,
              WindowManager.LayoutParams.FLAG_FULLSCREEN);
    }
    @Override
    protected void onResume() {
        super.onResume();
        // 判断显示哪个布局
        if (resumetag == 0)
            layout_welcome();
        else
            mainMenu();
    }
```

其中有一段代码,用来去除界面的标题,并且使整个界面占满屏幕,这段代码在这个项目中每个 Activity 都被用到:

```
requestWindowFeature(Window.FEATURE_NO_TITLE);
getWindow().setFlags(WindowManager.LayoutParams.FLAG_FULLSCREEN,
                WindowManager.LayoutParams.FLAG_FULLSCREEN);
```

其中 layout_welcome()显示的是 Logo 界面的方法,mainMenu()是显示主界面的方法。

在 layout_welcome()方法中,用户单击界面任意位置,改变了 resumetag 的值为 1,从而显示主界面,之所以在 OnResume()方法中调用而不是在 OnCreate()方法中调用,是因为 logo 界面只是在应用被加载的时候才显示,如果从其他页面跳回来到主界面,就不需要再显示了,layout_welcome()方法的代码如下:

```
    public void layout_welcome() {
        setContentView(R.layout.activity_welcome);// Logo 界面的布局文件
        ImageView welcomeImage;
        welcomeImage = (ImageView) findViewById(R.id.welcomeimage);
        welcomeImage.setOnClickListener(new OnClickListener() {
            @Override
```

```
        public void onClick(View v) {
            resumetag = 1;                          // 单击图片界面后将变量值改为1
            mainMenu();                             // 显示主界面
        }
    });
}
```

mainMenu()方法的代码如下：

```
public void mainMenu() {

        setContentView(R.layout.activity_mainmenu);
        xueqi = (TextView) findViewById(R.id.xueqiview);
        xuenian = (TextView) findViewById(R.id.xuenianview);
        gpa1 = (TextView) findViewById(R.id.gpa1view);
        gpa2 = (TextView) findViewById(R.id.gpa2view);
        pingjun = (TextView) findViewById(R.id.pingjunview);
        yihuo = (TextView) findViewById(R.id.yihuoview);
        SQLiteDatabase termdb = dbhelp.getReadableDatabase();
                Cursor cur = termdb.query(GPADatabaseHelper.
                                        TABLE_NAME_TERM, null, null, null, null, null,null);
        while (cur.moveToNext()) {
          xueqi.setText(Long.toString(cur.getLong(cur.getColumnIndex(GPAUsers.TERM))));
        } // 得到年份

        Calendar c = Calendar.getInstance();
        xuenian.setText(Integer.toString(c.get(Calendar.YEAR)));
        // 查询目前的数据结果
        // 常见 GPA
        gpa1.setText(new Calculate(MainActivity.this, 1, "hehao").returnStr);
        // 标准 GPA
        gpa2.setText(new Calculate(MainActivity.this, 2, "hehao").returnStr);
        // 平均分
        pingjun.setText(new Calculate(MainActivity.this, 3, "hehao").returnStr);
        // 已获学分
        yihuo.setText(new Calculate(MainActivity.this, 3, "hehao").SumCreditstr);
        exit = (ImageButton) findViewById(R.id.exitbutton);
        shurushuju = (Button) findViewById(R.id.inputbutton);
        chaxun = (ImageButton) findViewById(R.id.searchbutton);
        weilai = (ImageButton) findViewById(R.id.futurebutton);
        daochu = (ImageButton) findViewById(R.id.exportbutton);
        shezhi = (ImageButton) findViewById(R.id.settingbutton);
        exit.setOnClickListener(new buttonclicklistener());// 退出
        shurushuju.setOnClickListener(new buttonclicklistener());// 输入数据
```

```
        chaxun.setOnClickListener(new buttonclicklistener());// 查询
        weilai.setOnClickListener(new buttonclicklistener());// 未来
        daochu.setOnClickListener(new buttonclicklistener());// 导出
        shezhi.setOnClickListener(new buttonclicklistener());// 更多
}
```

(4) 这里有一个用于计算学分的类 Calculator.java：

```
public class Calculate {
    float Credit = 0;// 学分
    float SumCredit = 0;// 总学分
    float Achievement = 0;// 成绩
    float SumAchievement = 0;// 总成绩 * 学分

    public String returnStr;
    public String SumCreditstr;
    int Num = 0;
    float GPA;
    Context mcontext = null;

    /**
     *
     * @param context
     *            上下文对象
     * @param Ch
     *            选择哪种计算方法 1、2、3
     * @param usr
     *            用户的姓名
     */
    public Calculate(Context context, int Ch, String usr) {
        mcontext = context;
        GPADatabaseHelper gpahelper =
                new GPADatabaseHelper(mcontext,
                GPADatabaseHelper.TABLE_NAME_GPA, GPADatabaseHelper.VERSION);
        SQLiteDatabase gpaDb = gpahelper.getReadableDatabase();
        Cursor curcreateclass =
                        gpaDb.query(GPADatabaseHelper.TABLE_NAME_GPA,
                                null, GPAUsers.USERNAME + " = ?",
                        new String[] { usr }, null, null, GPAUsers._ID);
        switch (Ch) {
        case 1:// 方法一
            while (curcreateclass.moveToNext()) {
                Achievement = Float.parseFloat(curcreateclass
                    .getString(curcreateclass.getColumnIndex(GPAUsers.SCORE)));// 得到分数
```

```
            Credit = Float.parseFloat(curcreateclass
                    .getString(curcreateclass.getColumnIndex(GPAUsers.CREDIT)));
            SumCredit += Credit;
            if (Achievement < 60)// 分数在 60 以内
            Achievement = 0;
            else if (Achievement >= 60 && Achievement < 70)
                Achievement = 1;
            else if (Achievement >= 70 && Achievement < 80)
                Achievement = 2;
            else if (Achievement >= 80 && Achievement < 90)
                Achievement = 3;
            else if (Achievement >= 90 && Achievement <= 100)
                Achievement = 4;
                                        // 一共可以获得多少学分
            SumAchievement += Achievement * Credit;
        }
        break;
    case 2:// 方法二
        while (curcreateclass.moveToNext()) {
            Achievement = Float.parseFloat(curcreateclass
                    .getString(curcreateclass
.getColumnIndex(GPAUsers.SCORE)));
            Credit = Float.parseFloat(curcreateclass
                    .getString(curcreateclass
.getColumnIndex(GPAUsers.CREDIT)));
            SumCredit += Credit;
                                                // 得到学分乘以绩点
            SumAchievement += 4 * Achievement * Credit / 100;
        }
        break;
    case 3:// 方法三
        while (curcreateclass.moveToNext()) {
            Achievement = Float.parseFloat(curcreateclass.getString(
                                curcreateclass.getColumnIndex(GPAUsers.SCORE)));
            Credit = Float.parseFloat(curcreateclass.getString(
                                curcreateclass.getColumnIndex(GPAUsers.CREDIT)));
            SumCredit += Credit;
                                                // 得到学分乘以绩点
            SumAchievement += Achievement * Credit;
        }
        break;
    default:
```

```
                returnStr = "未找到所查用户";
                SumCreditstr = "";
                break;
            }
            if (Float.isNaN(SumAchievement / SumCredit))
                GPA = 0;
            else
                GPA = SumAchievement / SumCredit;
            SumCreditstr = Float.toString(SumCredit);
            returnStr = new DecimalFormat("0.00").format(GPA);
            curcreateclass.moveToFirst();
        }
    }
```

(5) 为 6 个按钮设置的监听事件是在 MainActivity.java 中建一个内部类：

```
class buttonclicklistener implements OnClickListener {
        @Override
        public void onClick(View v) {
            switch (v.getId()) {
            case R.id.exitbutton:// 退出
                Intent startMain = new Intent(Intent.ACTION_MAIN);
                startMain.addCategory(Intent.CATEGORY_HOME);
                startMain.setFlags(Intent.FLAG_ACTIVITY_NEW_TASK);
                startActivity(startMain);
                java.lang.System.exit(0);
                break;
            case R.id.inputbutton:// 输入数据
                Intent intent = new Intent();
                intent.putExtra("tag", 0);
                intent.putExtra("id", 0);
                intent.setClass(MainActivity.this, InfoinputActivity.class);
                startActivity(intent);
                break;
            case R.id.searchbutton:// 查询
                Intent intent1 = new Intent();
                intent1.setClass(MainActivity.this, ClassbrowerActivity.class);
                startActivity(intent1);
                break;
            case R.id.futurebutton:// 未来
                Intent intent2 = new Intent();
                intent2.setClass(MainActivity.this, FutureActivity.class);
                startActivity(intent2);
                break;
```

```
case R.id.exportbutton:// 导出
    try {
    doCopyFile();
    } catch (Exception e) {
      Toast.makeText(MainActivity.this,"请检查 sd 卡是否连接",
              Toast.LENGTH_LONG).show();
    e.printStackTrace();
    }
    break;
case R.id.settingbutton:// 更多
    Intent intent3 = new Intent();
    intent3.setClass(MainActivity.this, MoreActivity.class);
    startActivity(intent3);
    break;
}
}
}
```

单击“退出”按钮,则直接退出应用回到 HOME 界面,单击“输入数据”、“查询”、“未来”或“更多”按钮分别跳到相应界面,单击“导出”按钮,调用 doCopyFile()方法:

```
public void doCopyFile() throws Exception {
    File srcFile = new File("/data/data/ggsb.gpaclt/databases/GPAdata");
    File destFile = new File("/sdcard/chengji.sqlite");
    FileInputStream input = new FileInputStream(srcFile);

    try {
        FileOutputStream output = new FileOutputStream(destFile);
    try {
        byte[] buffer = new byte[4096];
        int n = 0;
        while (-1 != (n = input.read(buffer))) {
            output.write(buffer, 0, n);
        }
    } finally {
        try {
            if (output != null) {
                output.close();
            }
        } catch (IOException ioe) {
            ioe.printStackTrace();
        }
    }
```

```
        } finally {
            try {
                if (input != null) {
                    input.close();
                }
            } catch (IOException ioe) {
                ioe.printStackTrace();
            }
        }
        Toast.makeText(MainActivity.this,
                "已保存至 sd 卡根目录下,文件名为 chengji.sqlite",
                Toast.LENGTH_LONG).show();
    }
```

单击“输入数据”按钮时,跳到输入数据的界面,代码如下:

```
public class InfoinputActivity extends Activity {
    String[] termnum = { "1", "2", "3", "4", "5", "6", "7", "8" };// 学期
    private ArrayAdapter<String> adapter = null;
    private List<String> allterm = null;
    private EditText classnameedit = null;// 课程名称
    private Spinner termedit = null;// 开课学期
    private EditText scoreedit = null;// 成绩
    private EditText creditedit = null;// 学分
    private EditText otheredit = null;// 备注
    private RadioButton requiredCource = null;// 课程类型 必修
    private RadioButton elective = null;// 课程类型 选修
    private Button save1 = null;// 保存
    private Button save2 = null;// 存储
    private Button back = null;// 返回
    private Button continueinput = null;// 继续添加
    Intent intent = null;
    Long xueqi = null;
    Long termnumber = (long) 0;
    SQLiteDatabase termdb = null;
    Cursor cur = null;
        GPADatabaseHelper gpahelper =
                new GPADatabaseHelper(InfoinputActivity.this,
                GPADatabaseHelper.TABLE_NAME_GPA,
                GPADatabaseHelper.VERSION);
    String username = "hehao";
    String result = null;
    @Override
    protected void onCreate(Bundle savedInstanceState) {
```

```
        super.onCreate(savedInstanceState);
        inputlist();
}

public void inputlist() {
        requestWindowFeature(Window.FEATURE_NO_TITLE);
        getWindow().setFlags(WindowManager.LayoutParams.FLAG_FULLSCREEN,
                WindowManager.LayoutParams.FLAG_FULLSCREEN);
        setContentView(R.layout.activity_infoinput);

        //向数据库里存数据
        termdb = new GPADatabaseHelper(InfoinputActivity.this, "termdata", 1)
                .getWritableDatabase();
        cur = termdb.query("termdata", null, null, null, null, null, null);
        while (cur.moveToNext())
                {
            termnumber = cur.getLong(cur.getColumnIndex("term"));
        }
        classnameedit = (EditText) findViewById(R.id.classnameedit);
        termedit = (Spinner) findViewById(R.id.termedit);
        scoreedit = (EditText) findViewById(R.id.scoreedit);
        creditedit = (EditText) findViewById(R.id.creditedit);
        otheredit = (EditText) findViewById(R.id.othercontentedit);
        requiredCource = (RadioButton) findViewById(R.id.radio0);
        elective = (RadioButton) findViewById(R.id.radio1);
        save1 = (Button) findViewById(R.id.savebtn1);
        save2 = (Button) findViewById(R.id.savebtn2);
        back = (Button) findViewById(R.id.backbtn);
        continueinput = (Button) findViewById(R.id.ctnuinputbtn);

        allterm = new ArrayList<String>();
        for (int i = 0; i < termnum.length; i++)
            allterm.add(termnum[i]);
        adapter = new ArrayAdapter<String>(this,
              android.R.layout.simple_spinner_item, allterm);
                adapter.setDropDownViewResource(
                        android.R.layout.simple_spinner_dropdown_item);
        termedit.setAdapter(adapter);
        termedit.setOnItemSelectedListener(new Spinner.OnItemSelectedListener() {

            @Override
            public void onItemSelected(AdapterView<?> arg0, View arg1,
```

```
            int arg2, long arg3) {
        xueqi = arg3 + 1;
    }
    @Override
    public void onNothingSelected(AdapterView<? > arg0) {
    }
});
save1.setOnClickListener(new oklistener());
save2.setOnClickListener(new oklistener());
back.setOnClickListener(new oklistener());
continueinput.setOnClickListener(new oklistener());
}
```

单击按钮的监听事件是：

```
class oklistener implements OnClickListener {

    @Override
    public void onClick(View v) {
        switch (v.getId()) {
        case R.id.savebtn1:
            // 如果单击保存则没有 break 直接和存储实现同一个功能
            /*
             * 注：没有 break;
             */
        case R.id.savebtn2:// 保存之后关闭当前界面
            // 如果 check 返回值为 1 则表示成绩和学分符合要求
            if (check() == 1) {
                save();
                finish();
            }
            break;
              // 保存之后，将每个 EditText 设置为空，继续输入
        case R.id.ctnuinputbtn:
            if (check() == 1)
                save();
            classnameedit.setText("");
            scoreedit.setText("");
            creditedit.setText("");
            otheredit.setText("");
            break;
        case R.id.backbtn:// 关闭当前界面，返回上一界面
            finish();
            break;
```

```
        }
    }
```

其中的check()方法用于检测用户输入的成绩和学分是否正确：

```
/**
     * 检查文本框中成绩和学分是否输入正确
     *
     * @return tag 默认值为0,即输入得不正确
     *            输入正确时,将tag值改为1,并返回
     */
    public int check() {
        Float score = (float) 0;
        Float credit = (float) 0;
        int tag = 0;// 定义的变量,作为用户输入正确与否的标识
        while (tag == 0) {
            // 判断成绩是否为空或者学分是否为空
            if (TextUtils.isEmpty(scoreedit.getText().toString())
                    || TextUtils.isEmpty(creditedit.getText().toString())) {
        Toast.makeText(InfoinputActivity.this,"成绩、学分不能为空",
                        Toast.LENGTH_LONG).show();
                break;
            }
            // 判断成绩的小数点是否正确
            if (scoreedit.getText().charAt(0) != '.')
                // 正确的话转换成 float 类型的值
                    score = Float.parseFloat(scoreedit.getText().toString());
                else {
                    // 否则提示
                    Toast.makeText(InfoinputActivity.this,"成绩不能以小数点开头",
                            Toast.LENGTH_LONG).show();
                    break;
                }
                // 判断学分的小数点是否正确
                if (creditedit.getText().charAt(0) != '.')
                credit = Float.parseFloat(creditedit.getText().toString());
                else {
                    Toast.makeText(InfoinputActivity.this,"学分不能以小数点开头",
                            Toast.LENGTH_SHORT).show();
                    break;
                }
                // 判断分数是否在0～100之间
                if (score < 0) {
                    Toast.makeText(InfoinputActivity.this,"成绩不能小于0",
```

```
                Toast.LENGTH_SHORT).show();
                break;
            } else if (score>100) {
                Toast.makeText(InfoinputActivity.this,"成绩不能大于100",
                Toast.LENGTH_SHORT).show();
                break;
            }
            // 判断学分是否在0~10之间
            if (credit < 0) {
                Toast.makeText(InfoinputActivity.this,"学分不能小于0",
                Toast.LENGTH_SHORT).show();
                break;
            } else if (credit>10) {
            Toast.makeText(InfoinputActivity.this,"学分不能大于10",
            Toast.LENGTH_SHORT).show();
            break;
                }
                // 如果成绩和学分符合要求则将tag赋值为1
                tag = 1;
            }
            return tag;
        }
```

保存的时候调用的是save()方法,代码如下:

```
/**
 * 将写入的信息保存到数据库内
 */
public void save() {
    ContentValues classContent = new ContentValues();
    SQLiteDatabase okdatabase = gpahelper.getWritableDatabase();

    classContent.put(GPAUsers.USERNAME, username);
    classContent.put(GPAUsers.TERM, xueqi);
    if (xueqi>termnumber) {
        classContent.put(GPAUsers.TERM, xueqi);
    }
    cur.moveToFirst();
    if (cur.moveToNext() == true) {
        termdb.update(GPADatabaseHelper.TABLE_NAME_TERM,
                            classContent,
                GPAUsers._ID + " = ?",
                new String[] { Integer.toString(1) });
    } else
```

```
            termdb.insert(GPADatabaseHelper.TABLE_NAME_TERM, null,
                                        classContent);
        classContent.put(GPAUsers.CLASSNAME, classnameedit.getText()
                .toString());
        classContent.put(GPAUsers.SCORE,
                Float.parseFloat(scoreedit.getText().toString()));
        classContent.put(GPAUsers.CREDIT,
                Float.parseFloat(creditedit.getText().toString()));
        classContent.put(GPAUsers.CLASSTAG, requiredCource.isChecked() ? 1
            : 0);
        classContent.put(GPAUsers.OTHER, otheredit.getText().toString());
        if (intent.getIntExtra("tag", 0) == 0)
            okdatabase.insert(GPADatabaseHelper.TABLE_NAME_GPA,
                                        null,
                classContent);
        else
            okdatabase.update(GPADatabaseHelper.TABLE_NAME_GPA,
                    classContent, GPAUsers._ID + " = ?",
                    new String[] { Integer.toString(intent.getIntExtra("id", 0)) });
        classContent.put("tag", 1);
        result = new Calculate(InfoinputActivity.this, 1, username).returnStr;
    }
```

单击“保存”、“存储”或者“返回”的按钮之后，finish 当前的 Activity，回到 MainActivity 的界面。

单击“查询”按钮时，跳到查询数据的界面。

(6) 查询数据的界面类 ClassbrowerActivity.java，这个 Activity 比其他的 Activity 特别在，它要继承的不是 Activity，而是我们要接触的一个新的类，即 ListActivity，这个类是用于显示 ListView 的，代码如下：

```
public class ClassbrowerActivity extends ListActivity {
    /**
     * @see android.app.Activity#onCreate(Bundle)
     */
    int i = 0;
    String selection = null;
    String[] selectionarg = null;

    private Button back = null;// 返回
    private Button shaixuan = null;// 筛选
    private Button baocun = null;// 保存
    private Button xueqi = null;// 学期
    private Button leixing = null;// 类型
```

```
    private Button shanchu = null;// 删除

    public static int select_item = -1;
    SimpleAdapter menuAdapter = null;

    Cursor cur = null;

    GPADatabaseHelper classDbHelper = new GPADatabaseHelper(
            ClassbrowerActivity.this, GPADatabaseHelper.TABLE_NAME_GPA,
            GPADatabaseHelper.VERSION);

    //查询出的数据
    ArrayList<HashMap<String, String>> menuList = null;

    private static final String TAG = "classbrowerActivity";

    @Override
    protected void onCreate(Bundle savedInstanceState) {
        super.onCreate(savedInstanceState);
        layout();
    }

    public void layout() {
        requestWindowFeature(Window.FEATURE_NO_TITLE);
        getWindow().setFlags(WindowManager.LayoutParams.FLAG_FULLSCREEN,
                WindowManager.LayoutParams.FLAG_FULLSCREEN);
        setContentView(R.layout.activity_classbrowse);

        back = (Button) findViewById(R.id.backschpg);
        shaixuan = (Button) findViewById(R.id.choice);
        baocun = (Button) findViewById(R.id.saveschpg);
        xueqi = (Button) findViewById(R.id.xueqisx);
        leixing = (Button) findViewById(R.id.leixingsx);
        shanchu = (Button) findViewById(R.id.shanchusx);

        shaixuan.setOnClickListener(new buttonclicklistener());
        shanchu.setOnClickListener(new buttonclicklistener());
        leixing.setOnClickListener(new buttonclicklistener());
        back.setOnClickListener(new buttonclicklistener());
        baocun.setOnClickListener(new buttonclicklistener());
        xueqi.setOnClickListener(new buttonclicklistener());
    }
```

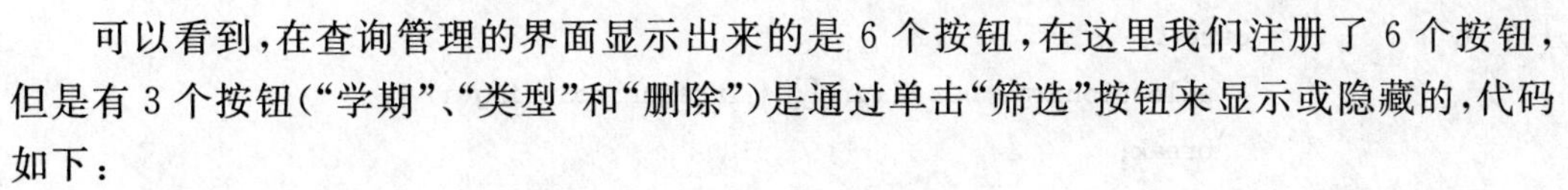

可以看到,在查询管理的界面显示出来的是 6 个按钮,在这里我们注册了 6 个按钮,但是有 3 个按钮("学期"、"类型"和"删除")是通过单击"筛选"按钮来显示或隐藏的,代码如下:

```
class buttonclicklistener implements OnClickListener {
    @Override
    public void onClick(View v) {
      switch (v.getId()) {
      case R.id.backschpg:// 返回
          finish();
          break;
        // 筛选,单击筛选的按钮是将类型、学期、删除3个按钮设为可见或不可见
        case R.id.choice:
          if (leixing.isShown()) {// ??
            leixing.setVisibility(View.INVISIBLE);
            xueqi.setVisibility(View.INVISIBLE);
            shanchu.setVisibility(View.INVISIBLE);
          } else {
            leixing.setVisibility(View.VISIBLE);
            xueqi.setVisibility(View.VISIBLE);
            shanchu.setVisibility(View.VISIBLE);
          }
          break;
        case R.id.saveschpg:// 新添加
            Intent intent = new Intent();
            intent.putExtra("tag", 0);
            intent.putExtra("id", 0);
            intent.setClass(ClassbrowerActivity.this,
                    InfoinputActivity.class);
            startActivity(intent);
            break;
          case R.id.xueqisx:// 学期
            selection = GPAUsers.TERM + " = ?";
            if (i < 8)
              i += 1;
            else
              i = 0;
            switch (i) {
            case 0:
              selection = null;
              selectionarg = null;
              break;
```

```
            case 1:
              selectionarg = new String[] { Integer.toString(i) };
              break;
            case 2:
              selectionarg = new String[] { Integer.toString(i) };
              break;
          case 3:
              selectionarg = new String[] { Integer.toString(i) };
              break;
          case 4:
              selectionarg = new String[] { Integer.toString(i) };
              break;
          case 5:
              selectionarg = new String[] { Integer.toString(i) };
              break;
          case 6:
              selectionarg = new String[] { Integer.toString(i) };
              break;
          case 7:
              selectionarg = new String[] { Integer.toString(i) };
              break;
          case 8:
              selectionarg = new String[] { Integer.toString(i) };
              break;
          }
          // 刷新界面
          onResume();
          break;
      case R.id.leixingsx:// 类型
          selection = GPAUsers.CLASSTAG + " = ?";

          if (i < 2)
            i += 1;
          else
            i = 0;
          switch (i) {
          case 0:
            selection = null;
            selectionarg = null;
            break;
          case 1:
            selectionarg = new String[] { Integer.toString(1) };
```

```
                break;
            case 2:
                selectionarg = new String[] { Integer.toString(0) };
                break;
            }
            onResume();
            break;
        case R.id.shanchusx:
            SQLiteDatabase deletedb =
                                classDbHelper.getWritableDatabase();
            deletedb.delete(GPADatabaseHelper.TABLE_NAME_GPA,
                                        GPAUsers.TAG
                + "= ?", new String[] { Integer.toString(0) });
            onResume();
            break;
        }
    }
}
```

(7) 查询界面要显示的内容需要通过适配器去显示,这里我们用到的是 SimpleAdapter,同时将代码放入到该 Activity 的 OnResume()方法中,便于每次刷新页面:

```
@Override
protected void onResume() {
    SQLiteDatabase db = classDbHelper.getWritableDatabase();
    ContentValues classContent = new ContentValues();
    classContent.put("tag", 1);
    db.update(GPADatabaseHelper.TABLE_NAME_GPA, classContent, null, null);
    cur = db.query(GPADatabaseHelper.TABLE_NAME_GPA, new String[] {
            GPAUsers._ID, GPAUsers.CLASSNAME, GPAUsers.CLASSTAG,
                        GPAUsers.TERM, GPAUsers.SCORE, GPAUsers.CREDIT,
                            GPAUsers.TAG },selection, selectionarg, null,
                                    null, GPAUsers.TERM + ", "
                                + GPAUsers.CLASSTAG + " DESC");

    //查询出来的课程的数据
    menuList = new ArrayList<HashMap<String, String>>();
    while (cur.moveToNext()) {
        //每项课程(包含各个属性)
        HashMap<String, String> Class = new HashMap<String, String>();
        //课程名称
        Class.put(GPAUsers.CLASSNAME,
                cur.getString(cur.getColumnIndex(GPAUsers.CLASSNAME)));
        //课程类型
```

```
            Class.put(GPAUsers.CLASSTAG, Integer.parseInt(cur.getString(
                        cur.getColumnIndex(GPAUsers.CLASSTAG))) == 1 ? "必修" : "选修");
            // 学期
            Class.put(GPAUsers.TERM, "第" +
                            cur.getString(cur.getColumnIndex(GPAUsers.TERM)) + "学期");
            // 成绩
            Class.put(GPAUsers.SCORE,
                    cur.getString(cur.getColumnIndex(GPAUsers.SCORE)));
            // 学分
            Class.put(GPAUsers.CREDIT,
                    cur.getString(cur.getColumnIndex(GPAUsers.CREDIT)));
            menuList.add(Class);
        }
    menuAdapter = new SimpleAdapter(this, menuList, R.layout.classmenu, new
                String[] { GPAUsers.CLASSNAME, GPAUsers.CLASSTAG,
              GPAUsers.TERM, GPAUsers.SCORE, GPAUsers.CREDIT },
    new int[] { R.id.lvclassname, R.id.lvleixing, R.id.lvterm,
            R.id.lvscore, R.id.lvcredit });
    setListAdapter(menuAdapter);
    getListView().setOnItemLongClickListener(new classlistlongclklistener());
    super.onResume();
}
```

数据显示后，单击 ListView 的数据时：

```
// 单击 ListView 的 item 时
    protected void onListItemClick(ListView l, View v, int position, long id) {
        super.onListItemClick(l, v, position, id);
        cur.moveToPosition(position);
        SQLiteDatabase updatedb = classDbHelper.getWritableDatabase();
        Cursor updcur = updatedb.query(GPADatabaseHelper.TABLE_NAME_GPA, null,
            GPAUsers._ID + " = ?",
            new String[] { Integer.toString(cur.getInt(0)) }, null, null,
            null);
        updcur.moveToNext();
        CheckedTextView clsnm = null;
        clsnm = (CheckedTextView) v.findViewById(R.id.lvclassname);
        clsnm.toggle();
        int tag = updcur.getInt(updcur.getColumnIndex("tag"));
        ContentValues classContent = new ContentValues();
        switch (tag) {
        case 1:
            classContent.put("tag", 0);
            break;
```

```
        case 0:
            classContent.put("tag", 1);
            break;
        }
        updatedb.update(GPADatabaseHelper.TABLE_NAME_GPA, classContent,
                GPAUsers._ID + " = ?",
                new String[] { Integer.toString(cur.getInt(0)) });
    }
```

（8）长按 ListView 中的数据时，跳到 InfoinputActivity 并通过 Intent 将数据传递过去：

```
// 长按适配器中的数据进入 infoinputActivity
class classlistlongclklistener implements OnItemLongClickListener {

    @Override
    public boolean onItemLongClick(AdapterView<? > arg0, View arg1,
            int position, long arg3) {
        cur.moveToPosition(position);
        Intent intent = new Intent();
        intent.putExtra("tag", 1);
        Log.d(TAG, "puttag = " + intent.getIntExtra("tag", 0));
        intent.putExtra("id", cur.getInt(0));
        Log.d(TAG, "putid = " + intent.getIntExtra("id", 0));
        intent.setClass(ClassbrowerActivity.this, InfoinputActivity.class);
        startActivity(intent);

        return false;
    }
}
```

InfoinputActivity 是如何接收的呢？我们需要在该 Activity 的 onCreate()方法中加入一段代码：

```
intent = getIntent();

if (intent.getIntExtra("tag", 0) == 1) {
    SQLiteDatabase gpadb = new GPADatabaseHelper(
        InfoinputActivity.this,
                GPADatabaseHelper.TABLE_NAME_GPA,
        GPADatabaseHelper.VERSION).getReadableDatabase();
    Cursor curclscnt =
            gpadb.query(GPADatabaseHelper.TABLE_NAME_GPA,
        null, GPAUsers._ID + " = ?", new String[] {
        Integer.toString(intent.getIntExtra("id", 0)) }, null,null, null);
```

```
        curclscnt.moveToNext();
        classnameedit.setText(curclscnt.getString(curclscnt
            .getColumnIndex(GPAUsers.CLASSNAME)));// 设置课程名称
        scoreedit.setText(curclscnt.getString(curclscnt
            .getColumnIndex(GPAUsers.SCORE)));// 成绩
        creditedit.setText(curclscnt.getString(curclscnt
            .getColumnIndex(GPAUsers.CREDIT)));// 学分
        otheredit.setText(curclscnt.getString(curclscnt
            .getColumnIndex(GPAUsers.OTHER)));// 备注
        if (curclscnt.getInt(curclscnt.getColumnIndex(GPAUsers.CLASSTAG)) == 1)
            requiredCource.setChecked(true);// 如果 == 1 必修
        else
            elective.setChecked(true);// 否则 选修
    }
```

(9) 单击返回按钮，返回到主界面，单击“未来”按钮，跳到 FutureActivity.java：

```
public class FutureActivity extends Activity {
    /**
     * @see android.app.Activity#onCreate(Bundle)
     */

    private Button futway1 = null;                // 方式一
    private Button futway2 = null;                // 方式二
    private Button calculate = null;              // 计算
    private Button futbackoff = null;             // 返回
    private Button modeintro = null;              // 方式说明
    private FrameLayout shuoming = null;
    private FrameLayout way = null;

    private EditText daihuoxuefen = null;         // 待获学分
    private EditText chengji = null;              // 所要达到的毕业平均成绩

    private TextView chengjifangshi = null;       // 成绩方式
    private TextView futureresult = null;         // 未来结果

    int tag = 0;

    @Override
    protected void onCreate(Bundle savedInstanceState) {
        super.onCreate(savedInstanceState);
        requestWindowFeature(Window.FEATURE_NO_TITLE);
        getWindow().setFlags(WindowManager.LayoutParams.FLAG_FULLSCREEN,
                WindowManager.LayoutParams.FLAG_FULLSCREEN);
```

```
        setContentView(R.layout.activity_future);

        futway1 = (Button) findViewById(R.id.futway1);
        futway2 = (Button) findViewById(R.id.futway2);
        calculate = (Button) findViewById(R.id.calculate);
        futbackoff = (Button) findViewById(R.id.backfutpg);
        modeintro = (Button) findViewById(R.id.modeintroduce);

        shuoming = (FrameLayout) findViewById(R.id.FrameLayout01);
        way = (FrameLayout) findViewById(R.id.way);

        daihuoxuefen = (EditText) findViewById(R.id.daihuoedit);
        chengji = (EditText) findViewById(R.id.chengjiedit);

        chengjifangshi = (TextView) findViewById(R.id.chengji);
        futureresult = (TextView) findViewById(R.id.futureresult);

        futway1.setOnClickListener(new buttonclicklistener());
        futway2.setOnClickListener(new buttonclicklistener());
        calculate.setOnClickListener(new buttonclicklistener());
        futbackoff.setOnClickListener(new buttonclicklistener());
        modeintro.setOnClickListener(new buttonclicklistener());
    }

    class buttonclicklistener implements OnClickListener {

        @Override
        public void onClick(View v) {
            switch (v.getId()) {
                // 单击方式一,将方式一的内容显示出来,其他组件不显示
            case R.id.futway1:/
                shuoming.setVisibility(View.INVISIBLE);
                way.setVisibility(View.VISIBLE);
                calculate.setVisibility(View.VISIBLE);
                futureresult.setVisibility(View.INVISIBLE);
                chengjifangshi.setText("所要达到的毕业平均成绩");
                tag = 1;
                break;
            case R.id.futway2:// 单击方式二,同方式一
                shuoming.setVisibility(View.INVISIBLE);
                way.setVisibility(View.VISIBLE);
                calculate.setVisibility(View.VISIBLE);
```

```
                futureresult.setVisibility(View.INVISIBLE);
                chengjifangshi.setText("今后计划达到的各科平均成绩");
                tag = 2;
                break;
            case R.id.modeintroduce:// 方式说明
                shuoming.setVisibility(View.VISIBLE);
                way.setVisibility(View.INVISIBLE);
                calculate.setVisibility(View.INVISIBLE);
                futureresult.setVisibility(View.INVISIBLE);
                break;
            case R.id.calculate:
                // 判断待获学分和成绩是否合法
                if (TextUtils.isEmpty(daihuoxuefen.getText().toString())
                    || TextUtils.isEmpty(chengji.getText().toString()))
                Toast.makeText(FutureActivity.this,"输入不能为空",
                        Toast.LENGTH_LONG).show();
                else if (daihuoxuefen.getText().charAt(0) == '.'
                        || chengji.getText().charAt(0) == '.')
                Toast.makeText(FutureActivity.this,"输入不能以小数点开头",
                        Toast.LENGTH_LONG).show();
                else {
                String suoqiu =
                        new Predict(Float.parseFloat(daihuoxuefen
                    .getText().toString()), Float.parseFloat(chengji
                        .getText().toString()), tag).resultstr;
                futureresult.setText(suoqiu);
                futureresult.setVisibility(View.VISIBLE);
                // Toast.makeText(futureActivity.this,suoqiu,
                // Toast.LENGTH_LONG).show();
            }
            break;
            case R.id.backfutpg:
            finish();
            break;
            }
        }
    }
```

单击"计算"按钮的时候,调用了一个我们创建的内部类:

```
/**
    * 计算以现在的成绩要想达到出国的水平还有多少差距
    * @author Administrator
    *
```

```
 */
public class Predict {
    Float // 已获总平均分
    SumAverAvhi = Float.parseFloat(new Calculate(FutureActivity.this, 3,
          "hehao").returnStr);
    Float // 已获学分
    GotCredit = Float.parseFloat(new Calculate(FutureActivity.this, 3,
          "hehao").SumCreditstr);
    Float MustBeAverAchi;// 今后要得的平均分
    String resultstr;

    public Predict(float FutureCredit, float AverAvhi, int tag) {
          // 方法一:输入目标平均分可获学分求要得的平均分
          switch (tag) {
          case 1:
              MustBeAverAchi = (AverAvhi * (GotCredit + FutureCredit)
              SumAverAvhi * GotCredit)/ FutureCredit;
              resultstr = "为达到该毕业成绩,剩余各科平均分应为"
                  + new  DecimalFormat("0.00").format(MustBeAverAchi);
              break;
            case 2:
              MustBeAverAchi = (SumAverAvhi * GotCredit + FutureCredit * AverAvhi)
                               / (GotCredit + FutureCredit);
              resultstr = "照此分数计算,毕业时获得的平均分应为"
                  + new DecimalFormat("0.00").format(MustBeAverAchi);
              break;
      }
    }
}
```

(10) 单击"更多"按钮,跳转到更多界面,即 MoreActivity.java:

```
public class MoreActivity extends Activity {
  /**
   * @see android.app.Activity#onCreate(Bundle)
   */
  private Button back = null;
  private Button about = null;
  private Button faq = null;
  private Button update = null;
  private Button product = null;
  String emailaddress = null;
  Intent intent = new Intent();
```

```
@Override
protected void onCreate(Bundle savedInstanceState) {
    requestWindowFeature(Window.FEATURE_NO_TITLE);
    getWindow().setFlags(WindowManager.LayoutParams.FLAG_FULLSCREEN,
        WindowManager.LayoutParams.FLAG_FULLSCREEN);
    super.onCreate(savedInstanceState);
    setContentView(R.layout.activity_more);

    back = (Button) findViewById(R.id.backmorepg);
    about = (Button) findViewById(R.id.aboutmorepg);// 关于
    faq = (Button) findViewById(R.id.faqmorepg);// FAQ
    update = (Button) findViewById(R.id.updatemorepg);// 更新
    product = (Button) findViewById(R.id.productmorepg);// 上传

    back.setOnClickListener(new buttonclicklistener());
    about.setOnClickListener(new buttonclicklistener());
    faq.setOnClickListener(new buttonclicklistener());
    // update.setOnClickListener(new buttonclicklistener());
    product.setOnClickListener(new buttonclicklistener());
}

class buttonclicklistener implements OnClickListener {

    @SuppressLint("WorldWriteableFiles")
    @SuppressWarnings("deprecation")
    @Override
    public void onClick(View v) {
        switch (v.getId()) {
        case R.id.backmorepg:// 返回按钮
          finish();
          break;
        case R.id.aboutmorepg:// 关于
          intent.setClass(MoreActivity.this, AboutActivity.class);
          startActivity(intent);
          break;
        case R.id.faqmorepg://"FAQ"按钮
          intent.setClass(MoreActivity.this, FaqlistActivity.class);
          startActivity(intent);
          break;
        // case R.id.updatemorepg:// 更新
        case R.id.productmorepg:// 上传
          final SharedPreferences emailsp =
```

```
                getSharedPreferences("email",
        Context.MODE_WORLD_WRITEABLE);
                                    // 自定义 Dialog 的视图
    final EditText input = new EditText(MoreActivity.this);
    emailaddress = emailsp.getString("email", null);
    input.setText(emailaddress);
    AlertDialog.Builder makesure = new AlertDialog.Builder(MoreActivity.this);
    makesure.setTitle("成绩上传")
  .setMessage("成绩将发送到邮箱:")
  .setCancelable(true)
  .setView(input)
  .setPositiveButton("确定",
new DialogInterface.OnClickListener() {

  @Override
  public void onClick(DialogInterface dialog, int which) {
      Editor editor = emailsp.edit();
      editor.putString("email", input.getText().toString());
      editor.commit();
      sendemail();
}
}).setNegativeButton("取消",new DialogInterface.OnClickListener() {

      @Override
      public void onClick(DialogInterface dialog, int which) {
      dialog.cancel();
      }
    }).show();
    break;
    }
  }
}
```

(11) 单击"关于"按钮的时候，跳到"关于"界面。"关于"界面是对项目的研发小组成员的介绍；单击"FAQ"按钮，进入 FAQ 界面，列出了 FAQ 的相关问题；"更新"按钮是应用升级时的更新；单击"上传"按钮，会弹出一个对话框，有输入邮箱的输入框、"确定"按钮和"取消"按钮，单击"确定"按钮就会将数据发送到指定的邮箱，单击"取消"按钮则会关闭对话框，实现上传功能的代码是内部的一个方法：

```
/**
   * 进行数据上传
   */
  void sendemail() {
```

```
        try {
            doCopyFile();// 将数据保存到 SDK 的方法
        } catch (Exception e) {
            Toast.makeText(MoreActivity.this,"请检查 sd 卡是否连接",
                    Toast.LENGTH_LONG) .show();
            e.printStackTrace();
        }
        File file = new File("/sdcard/chengji.sqlite"); // 附件文件地址
        Intent intent = new Intent(Intent.ACTION_SEND);
        String[] tos = { emailaddress };
        intent.putExtra(Intent.EXTRA_EMAIL, tos); //
        intent.putExtra(Intent.EXTRA_SUBJECT, file.getName()); //
        intent.putExtra(Intent.EXTRA_TEXT, "GPAdata-email sender"); // 正文
        // 添加附件,附件为 file 对象
        intent.putExtra(Intent.EXTRA_STREAM, Uri.fromFile(file));
        if (file.getName().endsWith(".gz")) {
            // 如果是 gz 使用 gzip 的 mime
            intent.setType("application/x-gzip");
            } else if (file.getName().endsWith(".txt")) {
              intent.setType("text/plain"); // 纯文本则用 text/plain 的 mime
            } else {
              // 其他的均使用流当做二进制数据来发送
              intent.setType("application/octet-stream");
            }
            startActivity(intent); // 调用系统的 mail 客户端进行发送
        }
```

这里我们调用了一个方法 doCopyFile(),这个方法我们在 MainActivity.java 中已经写好了,只需将代码复写到当前类中即可。

(12) 单击"关于"按钮,跳转到"关于"界面:

```
public class AboutActivity extends Activity {
    /**
     * @see android.app.Activity#onCreate(Bundle)
     */

    private Button back = null;

    @Override
    protected void onCreate(Bundle savedInstanceState) {
        super.onCreate(savedInstanceState);
        requestWindowFeature(Window.FEATURE_NO_TITLE);
        getWindow().setFlags(WindowManager.LayoutParams.FLAG_FULLSCREEN,
```

```
                WindowManager.LayoutParams.FLAG_FULLSCREEN);
        setContentView(R.layout.activity_about);

        back = (Button) findViewById(R.id.backabtpg);
        back.setOnClickListener(new clicklistener());
    }

    class clicklistener implements OnClickListener {

        @Override
        public void onClick(View v) {
            // switch (v.getId()) {
            // case R.id.backabtpg:
            finish();
            // break;
            // }
        }
    }
}
```

(13) 单击“FAQ”按钮,跳到 FAQ 界面,注意:这个 Activity 也继承了 ListActivity:

```
public class FaqlistActivity extends ListActivity {

    private Button back = null;

  String[] faqStrArray = { "1.什么是 GPA 成绩", "2.GPA 对于出国有什么作用", "3.美国的 GPA 算法是怎样的","4..举例显示", "5.英国的 GPA 成绩算法", "6.中文成绩单如何换算", "7.不同 GPA 算法如何取舍" };

    Intent intent = new Intent();

    @Override
    protected void onCreate(Bundle savedInstanceState) {
        requestWindowFeature(Window.FEATURE_NO_TITLE);
        getWindow().setFlags(WindowManager.LayoutParams.FLAG_FULLSCREEN,
                WindowManager.LayoutParams.FLAG_FULLSCREEN);
        super.onCreate(savedInstanceState);
        setContentView(R.layout.activity_faqlist);
        back = (Button) findViewById(R.id.backfaqpg);

        ArrayAdapter<? > adapter = new ArrayAdapter<String>(this,
                R.layout.textfaq, R.id.textinfaqlist, faqStrArray);
        setListAdapter(adapter);
```

```
        back.setOnClickListener(new buttonclicklistener());
    }

    class buttonclicklistener implements OnClickListener {

        @Override
        public void onClick(View v) {
            switch (v.getId()) {
            case R.id.backfaqpg:
                finish();
                break;
            }
        }
    }
```

(14) 当单击 ListView 中的数据时,通过 Intent 跳转到 FAQ 问题解答的界面,并把当前的问题传递过去,将问题和答案共同显示:

```
/**
 * 单击 ListView 中的数据,跳转到 FaqcontentActivity,并将被单击的数据的位置传递过去
 */
@Override
protected void onListItemClick(ListView l, View v, int position, long id) {
    super.onListItemClick(l, v, position, id);
    intent.putExtra("faqtag", position);
    intent.setClass(FaqlistActivity.this, FaqcontentActivity.class);
    startActivity(intent);
}
```

(15) 跳转到 FAQcontentActivity:

```
public class FaqcontentActivity extends Activity {
  /**
   * @see android.app.Activity#onCreate(Bundle)
   */

  private Button back = null;

  private RelativeLayout faq1 = null;
  private RelativeLayout faq2 = null;
  private RelativeLayout faq3 = null;
  private RelativeLayout faq4 = null;
  private RelativeLayout faq5 = null;
  private RelativeLayout faq6 = null;
```

```
private RelativeLayout faq7 = null;
ViewFlipper myflipper = null;

Intent intent = null;

int tag = 0;

GestureDetector det = null;

@Override
protected void onCreate(Bundle savedInstanceState) {
    requestWindowFeature(Window.FEATURE_NO_TITLE);
    getWindow().setFlags(WindowManager.LayoutParams.FLAG_FULLSCREEN,
          WindowManager.LayoutParams.FLAG_FULLSCREEN);
    super.onCreate(savedInstanceState);
    setContentView(R.layout.activity_faqcontent);

    myflipper = (ViewFlipper) this.findViewById(R.id.myFlipper);
    faq1 = (RelativeLayout) findViewById(R.id.faq1);
    faq2 = (RelativeLayout) findViewById(R.id.faq2);
    faq3 = (RelativeLayout) findViewById(R.id.faq3);
    faq4 = (RelativeLayout) findViewById(R.id.faq4);
    faq5 = (RelativeLayout) findViewById(R.id.faq5);
    faq6 = (RelativeLayout) findViewById(R.id.faq6);
    faq7 = (RelativeLayout) findViewById(R.id.faq7);

    intent = getIntent();
    tag = intent.getIntExtra("faqtag", 0);
    for (int i = tag; i>0; i--)
        myflipper.showNext();// ?

    back = (Button) findViewById(R.id.backfaqcontentpg);
    back.setOnClickListener(new buttonclicklistener());
    mygesture myg = new mygesture();
    det = new GestureDetector(FaqcontentActivity.this, myg);
}

class buttonclicklistener implements OnClickListener {

    @Override
```

```
            public void onClick(View v) {
                switch (v.getId()) {
                case R.id.backfaqcontentpg:
                    finish();
                    break;
                }
            }
        }
    class mygesture implements GestureDetector.OnGestureListener {

            @Override
            public boolean onDown(MotionEvent e) {
                return false;
            }

            @Override
            public boolean onFling(MotionEvent e1, MotionEvent e2, float velocityX,
                float velocityY) {
            if (e1.getX() - e2.getX()>0) {
                if (tag != 6) {
                    myflipper.setInAnimation(AnimationUtils.loadAnimation(
                                    FaqcontentActivity.this, R.anim.slide_in_right));
                    myflipper.setOutAnimation(AnimationUtils.loadAnimation(
                                    FaqcontentActivity.this, R.anim.slide_out_left));
                    myflipper.showNext();
                    tag++;
                    return true;
                    // intent.putExtra("faqtag", tag + 1);
                    // intent.setClass(FaqcontentActivity.this,
                    // FaqcontentActivity.class);
                    // finish();
                    // startActivity(intent);
                } else
                    ;
            }

            if (e1.getX() - e2.getX() < 0) {
            if (tag != 0) {
                myflipper.setInAnimation(AnimationUtils.loadAnimation(
                            FaqcontentActivity.this, R.anim.slide_in_left));
```

```
                myflipper.setOutAnimation(AnimationUtils.loadAnimation(
                            FaqcontentActivity.this, R.anim.slide_out_right));
                myflipper.showPrevious();
                  tag--;
                  return true;
                  // intent.putExtra("faqtag", tag-1);
                  // intent.setClass(FaqcontentActivity.this,
                  // FaqcontentActivity.class);
                  // finish();
                  // startActivity(intent);
            } else
                  ;
        }
        return false;
    }

    @Override
    public void onLongPress(MotionEvent e) {
    }

    @Override
    public boolean onScroll(MotionEvent e1, MotionEvent e2,
            float distanceX, float distanceY) {
        return false;
    }

    @Override
    public void onShowPress(MotionEvent e) {
    }

    @Override
    public boolean onSingleTapUp(MotionEvent e) {
        return false;
    }
}
    @Override
    public boolean onTouchEvent(MotionEvent event) {

        if (det.onTouchEvent(event))
            return det.onTouchEvent(event);
        else
            return super.onTouchEvent(event);
    }
```

6.6 本章小结

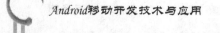

本章通过项目案例练习了几种布局的使用以及 Activity 之间的跳转和传值,介绍了两种适配器 ArrayAdapter 和 SimpleAdapter,这里需要理解的是适配器是数据和视图之间的桥梁,它的作用是将数据适配到视图当中进行显示。本章需要重点掌握 SQLite 的使用,包括数据库的创建以及增、删、改、查的方法,要理解 Cursor 的概念,它是数据库中每一行的集合,要通过列的名称和数据类型得到数据。同时注意 SQL 语句的正确使用,否则容易出现数据库无法正常操作的问题。

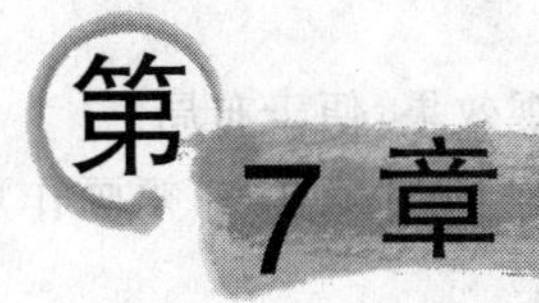

第7章 水墨丹青项目案例

水墨丹青(Ink Painter)是一款中国画绘画软件,它属于应用类软件。本章以水墨丹青项目为例介绍应用类软件项目 UI 的一般设计方法,以及本软件所涉及的绘图算法、毛笔仿真、数据处理、OpenGL ES 等相关内容。

7.1 预备知识

GLSurfaceView 是一个很好的基类对于构建一个使用 OpenGL ES 进行部分或全部渲染的应用程序。它的主要作用是:

- 提供黏合代码把 OpenGL ES 连接到视图系统。
- 提供黏合代码使得 OpenGL ES 按照 Acticity(活动)的生命周期工作。
- 使它容易选择一款合适的框架缓冲区像素格式。
- 创建和管理一个独立的渲染线程,产生平滑的动画。
- 提供更容易使用的调试工具来跟踪 OpenGL ES 的 API 并能找出错误。

渲染器是显卡中的元件,是 3D 引擎的核心部分,它完成将 3D 物体绘制到屏幕上的任务。它是一个公共接口,它的任务就是调用 OpenGL 的 API 来作帧的渲染。GLSurfaceView 的实现类通常会创建一个 Render 的实现类,然后用 setRenderer (GLSurfaceView. Renderer)方法把渲染器注册到 GLSurfaceView。

7.2 需求分析

本应用让用户体验到手指也能当毛笔,屏幕也能作宣纸的真实绘画感受,即手指在屏幕上绘出的笔画也会出现如同墨水在宣纸上晕开的独特效果。用户还能够实现毛笔蘸水墨色变淡、颜料混合出现新色、行笔变化"墨"趣横生等真实的绘画效果体验。

7.3 功能分析

本项目案例的功能包括以下几点。

(1) 毛笔墨迹仿真,用手指在屏幕上绘画能够体验到与毛笔在宣纸上绘画同样的感受。

(2) 可进行毛笔大小、颜色与墨色深浅的选择,并在窗口中呈现出当前画笔状态,方便用户选择。

(3) 通过菜单还可看到自己的画作装裱进画轴的远观效果,便于布局。

(4) 可实现最多 5 步的撤销与恢复,对画作可进行保存与读取,一幅画作可以多次完成。

7.4 设　计

水墨丹青软件的界面设计与功能描述如表 7-1 所示。

表 7-1　水墨丹青软件的界面设计与功能描述

界面	功能简述
主界面	• 显示绘制好的图画 • 调整画笔的大小 • 调整作图所用的颜料 • 绘制完成后保存 • 绘制完成后新建新的绘图界面 • 打开系统自带的图库,浏览保存的图片 • 关于产品的介绍 • 分享自己的作品 • 教你如何使用该产品
绘图界面	对图画的绘制,单击"Menu"键后可以返回主界面

7.4.1 UI 设计

(1) 水墨丹青软件开始画面如图 7-1 所示。

图 7-1　开始画面

(2) 主界面如图 7-2 所示。当用户单击主界面中央的画卷，进入绘图界面，如图 7-3 所示。(注:图 7-3 使用本软件绘制)绘画过程中按“Menu”键可返回主界面进行操作，作品会呈现在画轴中。单击不同大小的毛笔，能够实现不同类型毛笔的绘画特点。作品呈现如图 7-4 所示。

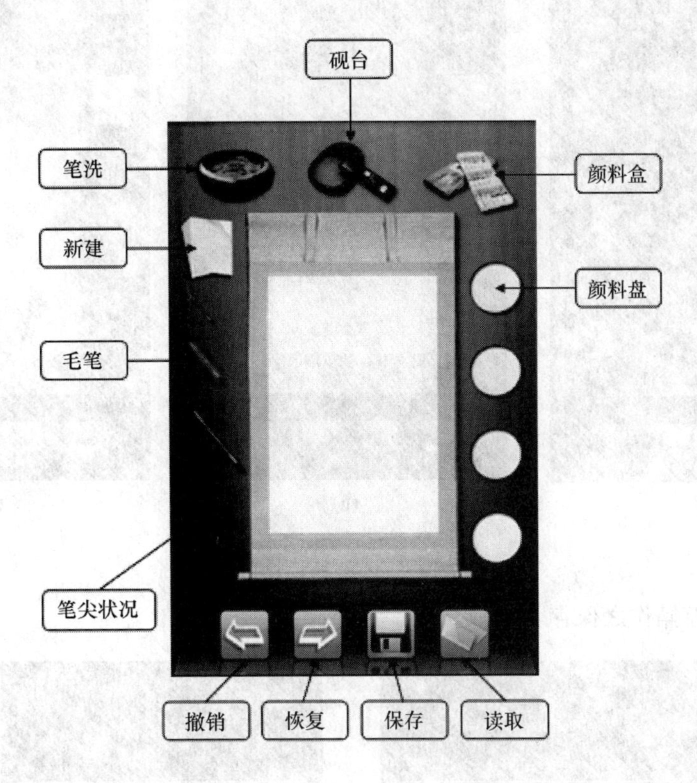

图 7-2 主界面

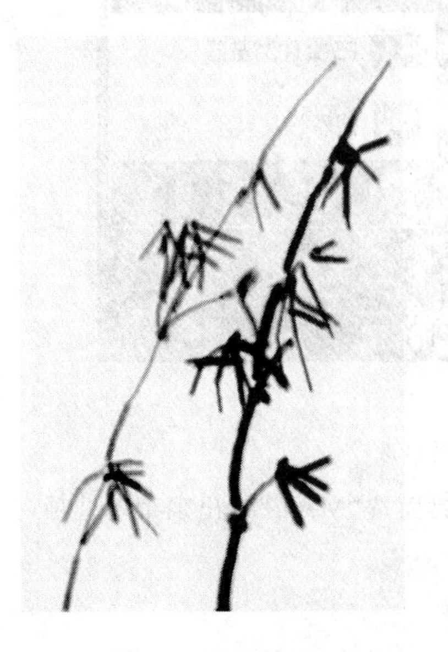

图 7-3 绘画界面

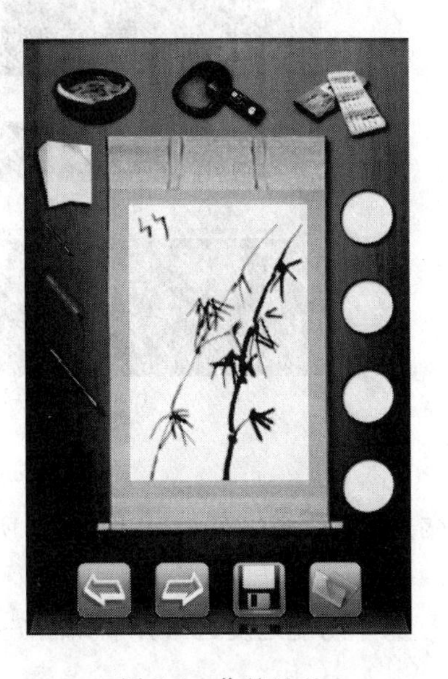

图 7-4 作品呈现

(3) 主界面操作之颜料选择如图 7-5 所示。第一步,单击颜料盒选择颜料;第二步,单击颜料盘,颜料会被挤入盘中。

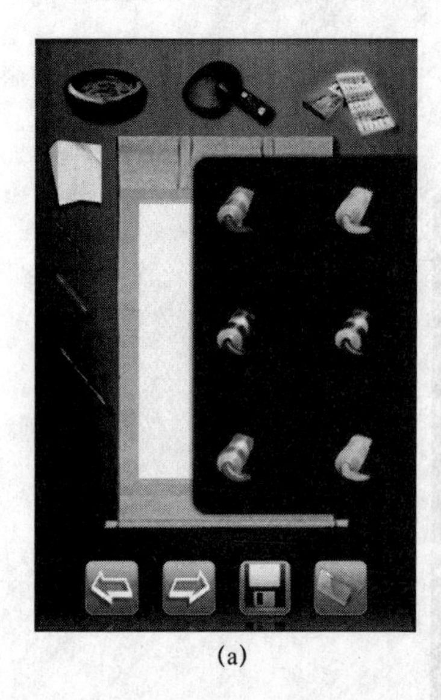

(a)

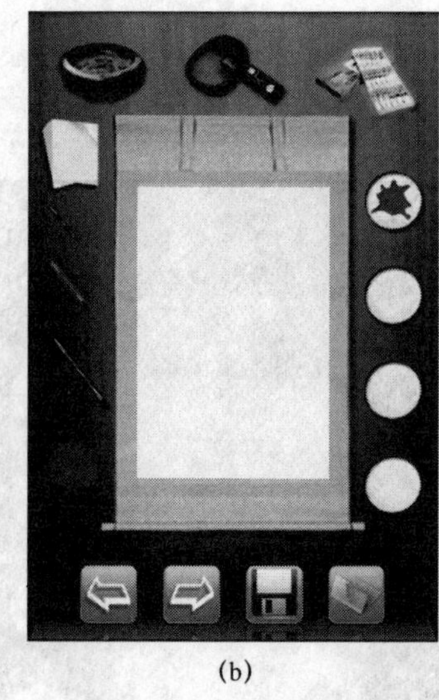

(b)

(c)

图 7-5　颜料选择

(4) 主界面操作之保存与读取如图 7-6 所示。

(a)

(b)

图 7-6　保存与读取

(5) 菜单功能如图 7-7～图 7-10 所示。主界面按"Menu"键出现菜单选项,如图 7-7 所示。

① 菜单功能之发送与分享如图 7-8 所示。

② 菜单功能之教学动画如图 7-9 所示。单击“教学”界面，会有生动的动画出现，向用户认真讲解“水墨丹青”的各项功能操作。“关于”界面如图 7-10 所示，该界面介绍了“水墨丹青”的版本号与它的制作团队，以及 Nermal 团队所要感谢的人。

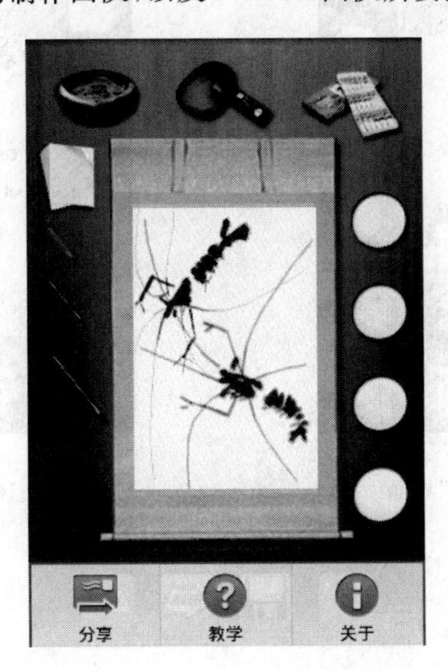

图 7-7　主菜单

图 7-8　发送与分享

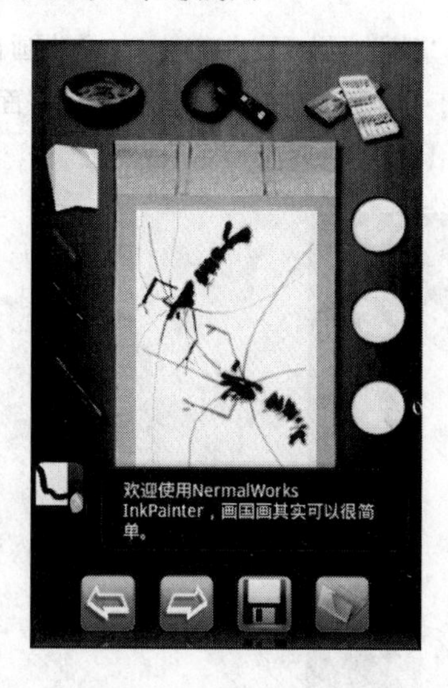

图 7-9 "教学"界面

图 7-10 "关于"界面

7.4.2 类设计

软件中用到的类如下：

- MenuTableAct.java。主视图，显示预览图和各种绘图工具，设置绘图参数。
- InkPainterAct.java。绘图视图，用于作画的"图纸"。
- GLRenderer.java。视图渲染器，画图的区域。

7.5 编程实现

MenuTableAct.java

(1) 代码中变量的声明

```
/** 毛笔的粗细 */
private final static int[][] BR_SIZE = {{2, 10}, {5, 20}, {10,35}};
private int m_WindowWidth;
private int m_WindowHeight;
//private int m_WindowWidth2;
//private int m_WindowHeight2;
/** 是否播放动画 */
//private boolean m_OnTurtorial = false;
//private boolean m_bOnAbout = false;
/** 3支毛笔的选中情况 */
private static int m_iBrush = 1;
/** 当前毛笔的颜色 */
private static int[] m_cBrush = {0xFF000000, 0xFF000000, 0xFF000000};
```

```
/** 当前毛笔的含水量
 * 0为全水,255为全墨/颜料
 **/
private static int[] m_wBrush = {0xFF, 0xFF, 0xFF};
/** 当前毛笔的干湿程度 */
//private static int[] m_BrushState = {BR_MILD, BR_MILD, BR_MILD};
/** 6种颜料选中情况,-1为未选中任何颜料 */
private static int m_iOil = -1;
/** 6种颜料的既定颜色 */
private final static int[] m_cOil =
{
        0xFFC92138, /* 红 */
        0xFFE2BA1C, /* 黄 */
        0xFF1E3E64, /* 蓝 */
        0xFF904211, /* 综 */
        0xFF53B491, /* 绿 */
        0xFF44C9F2, /* 天蓝 */
        0xFFFA8412  /* 橙色 */
};
/** 保存文件路径 存至SD卡 */
final public String Ink_PATH = "/sdcard/InkPainter/";
/** 保存的文件名 */
public String fileName;
/** 当前图片是否保存、修改 */
static boolean SaveRecord = false ;
/** 是否新建状态 */
public boolean NewBuild = false;
/** 读取图片 */
public Bitmap bitm ;
/** 是否读取状态 */
public boolean LoadBuild = false;
/** 是否分享状态 */
public boolean ShareBuild = false;
/** 盘子是否为空 */
private static boolean[] m_bDish = {false, false, false, false};
/** 4个盘子的颜色 */
private static int[] m_cDish = {0xFF000000, 0xFF000000, 0xFF000000, 0xFF000000};
/** 颜料混合系数 */
//private final static float OIL_ALPHA = 0.5f;
/** 是否使用颜料 */
```

```
private static boolean[] m_bInk = {true, true, true};
private ImageView m_Water, m_Ink, m_Color, m_NewDraw, m_Brush1, m_Brush2, m_Brush3;
public   ImageView m_Draw;
private ImageView m_UnDo, m_ReDo, m_Save, m_Load;
private ImageView m_UnDoShadow, m_ReDoShadow, m_SaveShadow, m_LoadShadow;
private ImageView m_Dish1, m_Dish2, m_Dish3, m_Dish4;
private ImageView m_Oil1, m_Oil2, m_Oil3, m_Oil4;
private ImageView m_Tip;
private ImageView m_OilSetting1, m_OilSetting2, m_OilSetting3, m_OilSetting4,
            m_OilSetting5, m_OilSetting6, m_OilSetting7, m_OilSetting8;
private PopupWindow m_Popup;
private ViewGroup m_ViewGroup;
private ScrollView m_Scroll;
private TimerTask m_TaskTutorial = null;
private TimerTask m_TaskAbout = null;
```

(2) 弹出窗口颜料颜色选择

使用8个ImageView来做窗口的颜料，每个ImageView的OnTouchListener事件为一种颜料的选择。

```
m_OilSetting1.setOnTouchListener(new OnTouchListener(){
  public boolean onTouch(View v, MotionEvent event){
    if(event.getAction() = = MotionEvent.ACTION_DOWN){
            m_OilSetting1.setImageResource(R.drawable.oilred2);
            m_OilSetting2.setImageResource(R.drawable.oilyellow);
            m_OilSetting3.setImageResource(R.drawable.oilblue);
            m_OilSetting4.setImageResource(R.drawable.oilbrown);
            m_OilSetting5.setImageResource(R.drawable.oildgreen);
            m_OilSetting6.setImageResource(R.drawable.oilsky);
            m_OilSetting7.setImageResource(R.drawable.oilorange);
            m_OilSetting8.setImageResource(R.drawable.cleaner);
    }else if(event.getAction() = = MotionEvent.ACTION_UP){
            //m_OilSetting1.setImageResource(R.drawable.oilred);
            //m_Popup.dismiss();
            m_iOil = 0;
    }
      return true;
  }});
```

(3) 蘸水，加墨，颜色的触发事件

蘸水的代码实现：

```
m_Water.setOnTouchListener(new OnTouchListener(){
  public boolean onTouch(View v, MotionEvent event){
```

```
        if(event.getAction() == MotionEvent.ACTION_DOWN){
        /* 结束"关于"动画 */
        StopAbout();
        /* 结束"教学"动画 */
        StopTutorial();
         m_Water.setImageResource(R.drawable.water2);
    }else if(event.getAction() == MotionEvent.ACTION_UP){
          m_Water.setImageResource(R.drawable.water);
          if (m_wBrush[m_iBrush] >= 10)
           m_wBrush[m_iBrush] -= 10;
           RefreshBrush();
        }
        return true;
    }
});
```

(4) 调整毛笔的大小

```
m_Brush1.setOnTouchListener(new OnTouchListener(){
    public boolean onTouch(View v, MotionEvent event){
          // TODO Auto-generated method stub
          if(event.getAction() == MotionEvent.ACTION_DOWN){
            /* 结束"关于"动画 */
           StopAbout();
           /* 结束"教学"动画 */
           StopTutorial();
           m_Brush1.setImageResource(R.drawable.brush2);
          }else if(event.getAction() == MotionEvent.ACTION_UP){
          m_iBrush = 0;
          RefreshBrush();
          }
          return true;
    }
});
```

(5)调色盘调整颜色

```
m_Dish1.setOnTouchListener(new OnTouchListener(){
    public boolean onTouch(View v, MotionEvent event){
          // TODO Auto-generated method stub
          if(event.getAction() == MotionEvent.ACTION_DOWN){
          /* 结束"关于"动画 */
          StopAbout();
          /* 结束"教学"动画 */
```

```
        StopTutorial();
        m_Dish1.setImageResource(R.drawable.pan2);
        /* 先判断是否选中颜料 */
        if (m_iOil >= 0){
            /* 是否清除颜料 */
            if (m_iOil == 7){
                /* 取消颜色 */
                m_bDish[0] = false;
                /* 取消颜料的选择 */
                m_iOil = -1;
                /* 去掉颜料 */
                m_Oil1.setImageResource(0);
                /* 恢复颜料盒图标 */
              m_Color.setImageResource(R.drawable.color);
        }else{
                /* 再判断盘中是否存在颜色 */
                if (m_bDish[0]){
              /* 调色 */
          m_cDish[0] = BlendColor(m_cDish[0], m_cOil[m_iOil]);
         }else{
              /* 加色 */
              m_cDish[0] = m_cOil[m_iOil];
              m_Oil1.setImageResource(R.drawable.oil1);
              m_bDish[0] = true;
              }
      /* 更新显示 */
        m_Oil1.setColorFilter(m_cDish[0],
      PorterDuff.Mode.MULTIPLY);
        /* 取消颜料的选择 */
        m_iOil = -1;
          /* 恢复颜料盒图标 */
      m_Color.setImageResource(R.drawable.color);
      }
  }
    /* 若没有选择颜料 */
  else{
      /* 先判断盘中是否存在颜色 */
      if (m_bDish[0]){
          /* 启用颜料 */
          m_bInk[m_iBrush] = false;
```

```
            /* 更新毛笔颜色 */
            m_cBrush[m_iBrush] = m_cDish[0];
            RefreshBrush();
          }
        }
      }else if(event.getAction() == MotionEvent.ACTION_UP){
            m_Dish1.setImageResource(R.drawable.pan);
      }
            return true;
      }
});
```

(6) 图片的保存

```
public void save(){
    LayoutInflater factory = LayoutInflater.from(MenuTable.this);
    final View v1 = factory.inflate(R.layout.dialogview,null);
    //R.layout.dialogview 与 dialogview.xml 文件名对应,把 login 转化成 View 类型
    AlertDialog.Builder dialog = new AlertDialog.Builder(MenuTable.this);
    dialog.setTitle(R.string.save_title);
    dialog.setIcon(R.drawable.icon);
    dialog.setView(v1);//设置使用 View
    //设置控件应该用 v1.findViewById,否则出错
    dialog.setPositiveButton(R.string.sure,new
    DialogInterface.OnClickListener() {
        public void onClick(DialogInterface dialog, int whichButton) {
        EditText text = (EditText)v1.findViewById(R.id.filename);
        fileName = text.getText().toString();
        try {
          saveFile(GLRenderer.m_bmpBackups[GLRenderer.GetBackupNum()],
                   fileName + ".png");
          if(SaveRecord){
               Toast.makeText(MenuTable.this,
               R.string.Save_succeed,Toast.LENGTH_SHORT).show();
               sendBroadcast(new Intent(Intent.ACTION_MEDIA_MOUNTED,
               Uri.parse("file://" + Environment.getExternalStorageDirectory())));
                if(NewBuild)
               {
                      GLRenderer.NewPaint();
                      m_Draw.setImageBitmap (GLRenderer.m_bmpBackups
                                          [GLRenderer.GetBackupNum()]);
                      SaveRecord = false ;
```

```
                    NewBuild = false;
                }
                if(LoadBuild)
                {
                    /* 将 Bitmap 设定到 ImageView */
                    m_Draw.setImageBitmap(bitm);
                   GLRenderer.m_bmpBackups[GLRenderer.GetBackupNum()] =
                                                              bitm;
                    GLRenderer.InitStack();
                    LoadBuild = false ;
                    //SaveRecord = true ;
                }
                if(ShareBuild)
                {
                    Uri uri = Uri.parse("file://" + Ink_PATH + fileName + ".png");
                    Intent intent = new Intent(Intent.ACTION_SEND);
                intent.setType("image/*");
                intent.putExtra(Intent.EXTRA_STREAM, uri);
                intent.putExtra(Intent.EXTRA_SUBJECT, "InkPainter");
                intent.putExtra (Intent.EXTRA_TEXT,
                            "This picture draw by InkPainter");
                startActivity(intent);//.createChooser(intent, getTitle()));
                ShareBuild = false ;
            }
        }
        else
          Toast.makeText(MenuTable.this,R.string.Save_fail,
                Toast.LENGTH_SHORT).show(); //不起作用
    }
    catch (IOException e)
    {
        // TODO Auto-generated catch block
        e.printStackTrace();
        Toast.makeText(MenuTable.this,R.string.Save_fail,
                Toast.LENGTH_SHORT).show();
    }
            }
        });
        dialog.setNegativeButton(R.string.cancel,new
  DialogInterface.OnClickListener() {
```

```
        @Override
        public void onClick(DialogInterface dialog, int which) {
            // TODO Auto-generated method stub
}
        });
        dialog.show();
    }
```

(7) 动画效果

```
public boolean onOptionsItemSelected(MenuItem item)
    {
        /* 结束"关于"动画 */
        StopAbout();
        /* 结束"教学"动画 */
        StopTutorial();
        switch (item.getItemId())
        {
            case 0:
                    if(! SaveRecord)
                    {
                        Toast.makeText(MenuTable.this,
                                R.string.Share_save, Toast.LENGTH_SHORT).show();
                        ShareBuild = true ;
                        save();
                    }
                    else
                    {
                        Uri uri = Uri.parse("file://" + Ink_PATH + fileName + ".png");
                        Intent intent = new Intent(Intent.ACTION_SEND);
                        intent.setType("image/*");
                        intent.putExtra(Intent.EXTRA_STREAM, uri);
                        intent.putExtra(Intent.EXTRA_SUBJECT, "InkPainter");
                        intent.putExtra(Intent.EXTRA_TEXT,
                                "This picture draw by InkPainter");
                        startActivity(intent);//.createChooser(intent, getTitle()));
                    }
                    return true;
                case 1:
                    /* 开始播放"教学"动画 */
                    {
```

```
                    if (m_TaskTutorial ! = null)
                    {
                        m_TaskTutorial.cancel();
                        m_TaskTutorial = null;
                    }
                        m_TaskTutorial = new TimerTask()
                        {
                            public void run()
                            {
                                Message message = new Message();
                                message.what = ID_TIMER_TUTORIAL;
                                m_Handler.sendMessage(message);
                            }
                        };
                    Timer timer = new Timer();
                    timer.schedule(m_TaskTutorial, 0, TUTORIAL_INTERVAL);
                    m_iTutorialStep = 0;
                }
                return true;
            case 2:
            /* 开始播放"关于"动画 */
            {
                    if (m_TaskAbout ! = null)
                    {
                        m_TaskAbout.cancel();
                        m_TaskAbout = null;
                    }
                        m_TaskAbout = new TimerTask()
                        {
                            public void run()
                            {
                                Message message = new Message();
                                message.what = ID_TIMER_ABOUT;
                                m_Handler.sendMessage(message);
                            }
                        };
                    Timer timer = new Timer();
                    timer.schedule(m_TaskAbout, 0, ABOUT_INTERVAL);
                    m_Scroll.setVisibility(View.VISIBLE);
                    m_Scroll.scrollTo(0, 0);
```

```
            }
            return true;
        }
        return false;
    }
/** Timer 相关 */
    /** 动画相关 */
    private final static int ID_TIMER_TUTORIAL = 411;
    private final static int TUTORIAL_INTERVAL = 2000;
    private final static int ID_TIMER_ABOUT = 1106;
    private final static int ABOUT_INTERVAL = 50;

    private int m_iTutorialStep = 0;
    private int m_iScrollPos = 0;
    Handler m_Handler = new Handler()
    {
            public void handleMessage(Message msg)
            {
                switch (msg.what)
                {
                /**"教学"动画 */
                case ID_TIMER_TUTORIAL:
                    switch (m_iTutorialStep)
                    {
                    case 0:
                        {
                            Toast toast =
            Toast.makeText(MenuTable.this,R.string.tutorial_0,
                                         Toast.LENGTH_LONG);
                            View view = toast.getView();
                            ImageView image = new ImageView(MenuTable.this);
                    LinearLayout lay =
                      new LinearLayout(MenuTable.this);
                      lay.setOrientation(LinearLayout.HORIZONTAL);
                      image.setImageResource(R.drawable.icon);
                      lay.addView(image);
                      lay.addView(view);
                      toast.setView(lay);
                      toast.show();
                        }
```

```
            break;
        case 1:
            {
                Toast toast =
                    Toast.makeText(MenuTable.this,
                    R.string.tutorial_1, Toast.LENGTH_LONG);
                View view = toast.getView();
                ImageView image = new ImageView(MenuTable.this);
                LinearLayout lay =
                        new LinearLayout(MenuTable.this);
                lay.setOrientation(LinearLayout.HORIZONTAL);
                image.setImageResource(R.drawable.icon);
                lay.addView(image);
                lay.addView(view);
                toast.setView(lay);
                toast.show();
            }
            break;
        case 2:
            m_iBrush = 0;
            RefreshBrush();
            break;
        case 3:
            m_iBrush = 1;
            RefreshBrush();
            break;
        case 4:
            m_iBrush = 2;
            RefreshBrush();
            break;
        case 5:
            {
                Toast toast =
                        Toast.makeText(MenuTable.this,
                        R.string.tutorial_2, Toast.LENGTH_LONG);
                View view = toast.getView();
                ImageView image = new ImageView(MenuTable.this);
                LinearLayout lay = new LinearLayout(MenuTable.this);
                lay.setOrientation(LinearLayout.HORIZONTAL);
                image.setImageResource(R.drawable.icon);
```

```
                    lay.addView(image);
                    lay.addView(view);
                    toast.setView(lay);
                    toast.show();
                }
                break;
            case 6:
                {
                    Toast toast =
                        Toast.makeText(MenuTable.this,
                        R.string.tutorial_3, Toast.LENGTH_LONG);
                    View view = toast.getView();
                    ImageView image = new ImageView(MenuTable.this);
                    LinearLayout lay = new LinearLayout(MenuTable.this);
                    lay.setOrientation(LinearLayout.HORIZONTAL);
                    image.setImageResource(R.drawable.icon);
                    lay.addView(image);
                    lay.addView(view);
                    toast.setView(lay);
                    toast.show();
                }
                break;
            case 7:
                m_Ink.setImageResource(R.drawable.yan2);
                break;
            case 8:
                m_Ink.setImageResource(R.drawable.yan);
                m_Water.setImageResource(R.drawable.water2);
                break;
            case 9:
              m_Water.setImageResource(R.drawable.water);
                  {
                   Toast toast =
                       Toast.makeText(MenuTable.this,
                       R.string.tutorial_4, Toast.LENGTH_LONG);
                    View view = toast.getView();
                    ImageView image = new ImageView(MenuTable.this);
                    LinearLayout lay = new LinearLayout(MenuTable.this);
                    lay.setOrientation(LinearLayout.HORIZONTAL);
                    image.setImageResource(R.drawable.icon);
```

```
                lay.addView(image);
                lay.addView(view);
                toast.setView(lay);
                toast.show();
                }
                break;
        case 10:
              m_Color.setImageResource(R.drawable.color2);
              break;
        case 11:
              m_Color.setImageResource(R.drawable.color);
              {
                  Toast toast =
                      Toast.makeText(MenuTable.this,
                      R.string.tutorial_5, Toast.LENGTH_LONG);
                  View view = toast.getView();
                  ImageView image = new ImageView(MenuTable.this);
                  LinearLayout lay = new LinearLayout(MenuTable.this);
                  lay.setOrientation(LinearLayout.HORIZONTAL);
                  image.setImageResource(R.drawable.icon);
                  lay.addView(image);
                  lay.addView(view);
                  toast.setView(lay);
                  toast.show();
                      }
                      break;
        case 13:
              {
                 Toast toast =
                     Toast.makeText(MenuTable.this,
                     R.string.tutorial_6, Toast.LENGTH_LONG);
                 View view = toast.getView();
                 ImageView image = new ImageView(MenuTable.this);
                 LinearLayout lay = new LinearLayout(MenuTable.this);
                 lay.setOrientation(LinearLayout.HORIZONTAL);
                 image.setImageResource(R.drawable.icon);
                 lay.addView(image);
                 lay.addView(view);
                 toast.setView(lay);
                 toast.show();
```

```
            }
        break;
    case 14:
        {
            Toast toast =
                Toast.makeText(MenuTable.this,
                R.string.tutorial_7, Toast.LENGTH_LONG);
            View view = toast.getView();
            ImageView image = new ImageView(MenuTable.this);
            LinearLayout lay = new LinearLayout(MenuTable.this);
            lay.setOrientation(LinearLayout.HORIZONTAL);
            image.setImageResource(R.drawable.icon);
            lay.addView(image);
            lay.addView(view);
            toast.setView(lay);
            toast.show();
            }
        break;
    case 15:
        m_UnDo.setImageResource(R.drawable.undo2);
        break;
    case 16:
        m_UnDo.setImageResource(R.drawable.undo);
        m_ReDo.setImageResource(R.drawable.redo2);
        break;
    case 17:
        m_ReDo.setImageResource(R.drawable.redo);
        {
            Toast toast =
                Toast.makeText(MenuTable.this,
                R.string.tutorial_8, Toast.LENGTH_LONG);
            View view = toast.getView();
            ImageView image = new ImageView(MenuTable.this);
            LinearLayout lay = new LinearLayout(MenuTable.this);
            lay.setOrientation(LinearLayout.HORIZONTAL);
            image.setImageResource(R.drawable.icon);
            lay.addView(image);
            lay.addView(view);
            toast.setView(lay);
            toast.show();
```

```
                }
            break;
        case 19:
            m_Save.setImageResource(R.drawable.save2);
            break;
        case 20:
            m_Save.setImageResource(R.drawable.save);
            m_Load.setImageResource(R.drawable.load2);
            break;
        case 21:
            m_Load.setImageResource(R.drawable.load);
            {
                Toast toast =
                    Toast.makeText(MenuTable.this,
                    R.string.tutorial_9, Toast.LENGTH_LONG);
                View view = toast.getView();
                ImageView image = new ImageView(MenuTable.this);
                LinearLayout lay = new LinearLayout(MenuTable.this);
                lay.setOrientation(LinearLayout.HORIZONTAL);
                image.setImageResource(R.drawable.icon);
                lay.addView(image);
                lay.addView(view);
                toast.setView(lay);
                toast.show();
                }
            break;
        case 22:
            m_Load.setImageResource(R.drawable.load);
        {
            Toast toast =
                Toast.makeText(MenuTable.this,
                R.string.tutorial_10, Toast.LENGTH_LONG);
            View view = toast.getView();
            ImageView image = new ImageView(MenuTable.this);
            LinearLayout lay = new LinearLayout(MenuTable.this);
            lay.setOrientation(LinearLayout.HORIZONTAL);
            image.setImageResource(R.drawable.icon);
            lay.addView(image);
            lay.addView(view);
            toast.setView(lay);
```

```
                    toast.show();
                }
            break;
        }
        m_iTutorialStep++;
        break;
    /** "关于"动画 */
    case ID_TIMER_ABOUT:
        m_iScrollPos = m_Scroll.getScrollY();
        m_iScrollPos += 1;
        m_Scroll.scrollTo(0, m_iScrollPos);
        break;
    }
    super.handleMessage(msg);
  }
};
/** 结束"关于"动画 */
private void StopAbout()
{
    if (m_TaskAbout != null)
    {
        m_TaskAbout.cancel();
        m_TaskAbout = null;
        m_Scroll.setVisibility(View.INVISIBLE);
    }
}
/** 结束"教学"动画 */
private void StopTutorial()
    {
    if (m_TaskTutorial != null)
    {
        m_TaskTutorial.cancel();
        m_TaskTutorial = null;
    }
}
```

(8) 图片的读取和保存

```
/** 读取图片 */
@Override
protected void onActivityResult(int requestCode,
                                int resultCode, Intent data)
```

```
{
  if (resultCode == RESULT_OK)
  {
    Uri uri = data.getData();
    String fName = uri.getPath();
    File f = new File(fName);

   //String end =
fName.substring(fName.lastIndexOf(".") + 1,
                              fName.length()).toLowerCase();
    //if(end.equals("png"))
    //{
ContentResolver cr = this.getContentResolver();
      BitmapFactory.Options opts = new BitmapFactory.Options();
      if (f.length() < 20480) { // 0 - 20k
          opts.inSampleSize = 1;
        } else if (f.length() < 51200) { // 20 - 50k
          opts.inSampleSize = 2;
        } else if (f.length() < 307200) { // 50 - 300k
          opts.inSampleSize = 4;
        } else if (f.length() < 819200) { // 300 - 800k
          opts.inSampleSize = 6;
        } else if (f.length() < 1048576) { // 800 - 1024k
          opts.inSampleSize = 8;
        } else {
          opts.inSampleSize = 10;
        }

      try
      {
      Bitmap bitmap =
          BitmapFactory.decodeStream(cr .openInputStream(uri),
                                  null, opts);
          //Bitmap bitmap = BitmapFactory.decodeFile(uri.getPath());
bitm = Bitmap.createScaledBitmap(bitmap,
                 m_WindowWidth, m_WindowHeight,true);
          new AlertDialog.Builder(MenuTable.this)
              .setTitle(R.string.Load_text)
              .setIcon(R.drawable.icon)
              .setPositiveButton
```

```
(
R.string.sure,
new DialogInterface.OnClickListener()
{
    @Override
    public void onClick(DialogInterface arg0, int i)
    {
        if(SaveRecord)
        {
            /* 将 Bitmap 设定到 ImageView */
            m_Draw.setImageBitmap(bitm);
//int data[] = new int[m_WindowWidth * m_WindowHeight];
        //bitm.getPixels(data, 0, m_WindowWidth,
    0, 0, m_WindowWidth, m_WindowHeight);
        GLRenderer.InitStack();
GLRenderer.m_bmpBackups[GLRenderer.GetBackupNum()].setPixels(data,
    m_WindowWidth * m_WindowHeight-m_WindowWidth,
    -m_WindowWidth, 0, 0, m_WindowWidth, m_WindowHeight);
GLRenderer.m_bmpBackups[GLRenderer.GetBackupNum()] = bitm;
        }
        else
        {
            new AlertDialog.Builder(MenuTable.this)
            .setTitle(R.string.Load_save_text)
            .setIcon(R.drawable.icon)
            .setPositiveButton
            (
            R.string.yes,
            new DialogInterface.OnClickListener()
            {
        @Override
        public void onClick(DialogInterface arg0, int i)
        {
            LoadBuild = true;
              save();
        }
    }
    )
    .setNegativeButton
    (
```

```
                    R.string.no,
                    new DialogInterface.OnClickListener()
                    {
                        @Override
                        public void onClick(DialogInterface arg0, int i)
                        {
                            /* 将 Bitmap 设定到 ImageView */
                            m_Draw.setImageBitmap(bitm);
                    //int data[] =
        new int[m_WindowWidth * m_WindowHeight];
                    /* bitm.getPixels(data, 0, m_WindowWidth, 0, 0, m_WindowWidth,
m_WindowHeight); */
                                GLRenderer.InitStack();
/* GLRenderer.m_bmpBackups[GLRenderer.GetBackupNum()].setPixels(data,
m_WindowWidth * m_WindowHeight-m_WindowWidth,-m_WindowWidth, 0, 0,
m_WindowWidth, m_WindowHeight); */
    GLRenderer.m_bmpBackups[GLRenderer.GetBackupNum()] = bitm;
                                }
                            }
                            ).show();
                        }
                    }
                }
                )
                .setNegativeButton
                (
                R.string.cancel,
                new DialogInterface.OnClickListener()
                {
                    @Override
                    public void onClick(DialogInterface arg0, int i)
                    {
                    }
                }
                ).show();
                }
                catch (FileNotFoundException e)
                {
                    e.printStackTrace();
                }
```

```
        }
    // else
    //      Toast.makeText(MenuTable.this,
                fName, Toast.LENGTH_LONG).show();
    // }
    super.onActivityResult(requestCode, resultCode, data);
    }
/** 保存 **/
  public void saveFile(Bitmap bm, String fileName) throws IOException
  {
      File dirFile = new File(Ink_PATH);
      if(! dirFile.exists())
      {
          dirFile.mkdir();
      }
      FileOutputStream fos = new FileOutputStream(Ink_PATH + fileName);
      bm.compress(Bitmap.CompressFormat.PNG, 100, fos);
      fos.flush();
      fos.close();
      //Log.v("0","0");
      SaveRecord = true ;
  }
```

(9) InkPainter.java 类中的变量的声明

```
private GLSurfaceView m_Surface;
GLRenderer m_Renderer;
private boolean m_bPopUpMenu = false;
private static boolean m_exit = false;
```

(10) 全屏、视图、菜单的设置

```
public void onCreate(Bundle savedInstanceState) {
        super.onCreate(savedInstanceState);
        /* 全屏显示 */
          requestWindowFeature(Window.FEATURE_NO_TITLE);
          getWindow().setFlags(WindowManager.LayoutParams.FLAG_FULLSCREEN,
                  WindowManager.LayoutParams.FLAG_FULLSCREEN);
        /* 将 GLRenderer 设为当前 View */
          m_Surface = new GLSurfaceView(this);
          m_Renderer = new GLRenderer(this);
          m_Surface.setRenderer(m_Renderer);
          setContentView(m_Surface);
        /* 最先进入开场动画 */
```

```
        /* 再进入 Menu 菜单 */
        Intent intent = new Intent();
            intent.setClass(InkPainter.this, Splash.class);
            intent.setFlags(Intent.FLAG_ACTIVITY_CLEAR_TOP);
            startActivity(intent);
              /* 启动处理机制 */
m_TimerDraw.schedule(m_Task, 0, PROCESS_INTERVAL);
// Log.v("Create","Create");
}
```

(11) 绘图的制作

```
public boolean onTouchEvent(MotionEvent event) {
        int x, y;
        x = (int) event.getX();
        y = (int) event.getY();
        switch (event.getAction()) {
        /* 单击屏幕 */
        case MotionEvent.ACTION_DOWN:
            m_Renderer.TouchDown(x, y);
            MenuTable.SaveRecord = false;
            // m_Renderer.TestPoint((int)x, (int)y);
            break;
        /* 拖动 */
        case MotionEvent.ACTION_MOVE:
            m_Renderer.AddParticles(x, y);
            break;
        /* 离开屏幕 */
        case MotionEvent.ACTION_UP:
          m_Renderer.ScreenShot();
            m_Renderer.TouchUp();
            break;
        }
        return true;
    }
  /** Timer 相关 */
    private final static int ID_TIMER_PROCESS = 711;
    private final static int PROCESS_INTERVAL = 30;
    Timer m_TimerDraw = new Timer();
    Handler m_Handler = new Handler() {
        public void handleMessage(Message msg) {
            switch (msg.what) {
```

```
            case ID_TIMER_PROCESS:
                if (m_exit) {
                    finish();
                    System.exit(0);
                }
                /* 惯性衰减 */
                m_Renderer.DecayParticles();
                /* 粒子扩散 */
                m_Renderer.SpreadParticles();
                /* 弹出菜单 */
                if (m_bPopUpMenu) {
                /* 判断是否成功截屏 */
                if (! m_Renderer.IsScreenShot()) {
                    Intent intent = new Intent();
                    intent.setClass(InkPainter.this, MenuTable.class);
                    intent.setFlags(Intent.FLAG_ACTIVITY_CLEAR_TOP);
                    startActivity(intent);
                    overridePendingTransition(R.anim.zoomin, 0);
                m_bPopUpMenu = false;
                }
            }
        break;
        }
        super.handleMessage(msg);
    }
};
TimerTask m_Task = new TimerTask() {
    public void run() {
        Message message = new Message();
        message.what = ID_TIMER_PROCESS;
        m_Handler.sendMessage(message);
    }
};
```

(12) GLRenderer 中变量的声明

```
/*
 ******************常量区域******************
 */
 /** 纹理区域坐标 */
 private final static float[] TEXTURE_COORDS =
     new float[]
```

```
                {
              1, 1,
              0, 1,
              0, 0,
              1, 0
                };
  /** 背景顶点坐标 */
  private final static float[] BACK_COORDS =
      new float[]
                {
              1f, 1f,
              -1f, 1f,
              -1f, -1f,
              1f, -1f,
              };
  /** 循环队列能容纳最大粒子数 */
  private final static int MAX_PARTICLE = 5000;
  /** 有多少死去的粒子作为存活粒子处理 */
  private final static int DEAD_RESERVE = 1000;
  /** 最大能够撤销的步数 */
  private final static int MAX_BACKUP = 7;
  /*
  *******************变量区域*******************
   */
  /** 纹理最大尺寸,构造函数中根据屏幕大小决定 */
  private static int MAX_TEXTURE_SIZE;
  /** 纹理大小 */
  private static int TEXTURE_SIZE;
  /** 屏幕大小 */
  private static int SCREEN_SIZE;
  /** 记录上下文 */
  private Context m_Context;
  /** 记录顶点与纹理的标准坐标 */
  private static FloatBuffer m_fTextureBuf;
  private static FloatBuffer m_fBackBuf;
  private static FloatBuffer m_fBackTexBuf;
  /** 纹理索引 */
  private IntBuffer m_iTextureBuf;
  /** 记录窗体大小 */
  private static int m_WindowWidth, m_WindowHeight;
```

```
/** 简化运算,窗体大小一半 */
private int m_WindowWidth2, m_WindowHeight2;
/** 粒子 */
private Particle[] m_Particles = new Particle[MAX_PARTICLE];
/** 最年轻粒子编号 */
private int m_ipTail;
/** 最年老粒子编号 */
private int m_ipHead;
/** 尸骨未寒粒子起始编号 */
private int m_ipCadaver;
/** 此时新生粒子标准尺寸,为零则表示当前没有墨水沾到纸上 */
private float m_fCurSize;
/** 此时新生粒子标准颜色 */
private static float[] m_Color = new float[3];
/** 记录光标(接触点)位置 */
private int m_CursorX, m_CursorY;
/** 是否在触摸中 */
private boolean m_bTouch;

/** 是否进行更新操作 */
private static boolean m_bUpdate;
/** 是否进行截屏操作 */
private boolean m_bScreenShot;
/** 屏幕备份 */
public static Bitmap[] m_bmpBackups = new Bitmap[MAX_BACKUP];
/** 屏幕备份序号 */
private static int m_iBackup;
/** 栈顶栈底指针 */
private static int m_ipTop;
private static int m_ipBottom;
/** 当前画笔粗细最大值 */
private static int m_iBrushMax;
/** 当前画笔粗细最小值 */
private static int m_iBrushMin;
/** 当前笔上墨水含量(纹理的 Alpha 值) */
private static float m_fInkAmount;
/** 是墨水还是颜料 */
private static boolean m_bInk = true;
/** 纹理数据 */
//private int[] m_TextureData;
```

```
/** 屏幕数据 */
//private int[] m_ScreenData;
/** 纹理位图 */
//private Bitmap m_bmpTexture;
```

(13) 粒子的相关事件

```
/** 添加粒子
  * 画笔移动时触发 */
  public void AddParticles(int x, int y)
  {
      /* 先判断当前是否有墨水沾到纸上 */
      if (m_fCurSize>0)
      {
          /* 采用 Bresenham 算法插值 */
          int   i, p, xcur = m_CursorX, ycur = m_CursorY;
          int   dx = x-m_CursorX;
          int   dy = y-m_CursorY;
          int  dx2; // 2 * dx
          int  dy2; // 2 * dy
        int  xi; // x 轴增量
          int  yi; // y 轴增量
        int  error; // 与真实直线误差
        /* 转换坐标为第一象限 */
          if (dx >= 0)
              xi = 1;
          else
          {
              dx = - dx;
              xi = - 1;
          }
        if (dy >= 0)
              yi = 1;
        else
        {
              dy = - dy;
              yi = - 1;
        }
      /* 简化运算 */
        dx2 = dx << 2;
        dy2 = dy << 2;
```

```
/* 根据直线是近 x 轴线还是远 x 轴线插值 */
if (dx>dy) // 若为近 x 轴线
{
    /* 初始化起始误差 */
    error = dy2 - dx;
    /* 忽略起点 */
    xcur += xi;
    for (i = 0; i < dx; i++)
    {
        /* 插值 */
        /* 循环队列 */
        p = m_ipTail % MAX_PARTICLE;
        m_Particles[p].x = xcur / (float)m_WindowWidth2 - 1f;
        m_Particles[p].y = 1f - ycur / (float)m_WindowHeight2;
        m_Particles[p].size = m_fCurSize;
        /* 颜色 */
        m_Particles[p].color[0] = m_Color[0];
        m_Particles[p].color[1] = m_Color[1];
        m_Particles[p].color[2] = m_Color[2];
        /* 生命 */
        m_Particles[p].life = 20;
        /* 墨量 */
        m_Particles[p].ink = m_fInkAmount;
        /* 移动队尾 */
        m_ipTail++;
        /* 衰减墨量 */
        //m_fInkAmount -= 0.0005f;
        /* 修正误差 */
        if (error >= 0)
        {
            ycur += yi;
            error -= dx2;
        }
        xcur += xi;
        error += dy2;
        //Log.v("xcur", Integer.toString(xcur));
        //Log.v("ycur", Integer.toString(ycur));
    }
    //Log.v("近 x", "近 x");
}
```

```
        else // 若为远 x 轴线
        {
            /* 初始化起始误差 */
            error = dx2 - dy;
            /* 忽略起点 */
            ycur += yi;
            for (i = 0; i < dy; i ++ )
            {
                /* 插值 */
                /* 循环队列 */
                p = m_ipTail % MAX_PARTICLE;
                m_Particles[p].x = xcur / (float)m_WindowWidth2 - 1f;
                m_Particles[p].y = 1f - ycur / (float)m_WindowHeight2;
                m_Particles[p].size = m_fCurSize;
                /* 颜色 */
                m_Particles[p].color[0] = m_Color[0];
                m_Particles[p].color[1] = m_Color[1];
                m_Particles[p].color[2] = m_Color[2];
                /* 生命 */
                m_Particles[p].life = 20;
                /* 墨量 */
                m_Particles[p].ink = m_fInkAmount;
                /* 移动队尾 */
                m_ipTail ++ ;
                /* 衰减墨量 */
                //m_fInkAmount -= 0.0005f;
                /* 修正误差 */
                if (error >= 0)
                {
                    xcur += xi;
                    error -= dy2;
                }
                ycur += yi;
                error += dx2;
                //Log.v("xcur", Integer.toString(xcur));
                //Log.v("ycur", Integer.toString(ycur));
            }
            //Log.v("远 x", "远 x");
        }
    if (m_bTouch)
```

```
        {
            float l = (float)Math.sqrt(dx * dx + dy * dy);
            //Log.v("l", Float.toString(l));
            /* 根据移动速度调整笔迹粗细 */
            m_fCurSize -= (l - 5) / 8;
            if (m_fCurSize>m_iBrushMax)
                    m_fCurSize = m_iBrushMax;
            else if (m_fCurSize < m_iBrushMin)
                    m_fCurSize = m_iBrushMin;
            }
        }
        m_CursorX = x;
        m_CursorY = y;
    }
    /** 粒子的惯性衰减 */
    public void DecayParticles()
    {
        if (! m_bTouch)
        {
            m_fCurSize -= (1 + 0.3 * m_fCurSize);
            if (m_fCurSize < 0)
                m_fCurSize = 0;
        }
    }
    /** 粒子扩散效应 */
    public void SpreadParticles()
    {
        int i, p;
        for (i = m_ipHead; i < m_ipTail; i++)
        {
            p = i % MAX_PARTICLE;
            m_Particles[p].size *= (1.01 + (Math.random() - 0.5) * 0.01);
            m_Particles[p].color[0] -= 0.001;
            m_Particles[p].color[1] -= 0.001;
            m_Particles[p].color[2] -= 0.001;
            m_Particles[p].ink -= 0.001;
            m_Particles[p].life--;
            if (m_Particles[p].life <= 0)
                m_ipHead = i + 1;
        }
```

```
        }
```

(14) 纹理的处理

```
/** 重置全部粒子
     * 笔尖按下时调用
     */
    private void ResetParticles()
    {
        m_fCurSize = m_iBrushMin + 1;
    }
    /** 读取纹理 */
    private void LoadTextures(GL10 gl)
    {
        /* 为纹理索引缓存分配空间
         * 0为粒子,1为背景,2为背景备份 */
        gl.glEnable(GL10.GL_TEXTURE_2D);

        /* 若为第一次装载纹理,则从资源文件中装载粒子纹理和背景纹理  */
        m_iTextureBuf = IntBuffer.allocate(3);
        gl.glGenTextures(3, m_iTextureBuf);
        /* 从资源中读取纹理图片 */
        Bitmap texture =
              Utils.getTextureFromBitmapResource(m_Context, R.drawable.particle);
        m_fBackTexBuf = Utils.newFloatBuffer(8);
        m_fBackTexBuf.put(0, (float)m_WindowWidth / MAX_TEXTURE_SIZE);
        m_fBackTexBuf.put(1, (float)m_WindowHeight / MAX_TEXTURE_SIZE);
        m_fBackTexBuf.put(2, 0f);
        m_fBackTexBuf.put(3, (float)m_WindowHeight / MAX_TEXTURE_SIZE);
        m_fBackTexBuf.put(4, 0f);
        m_fBackTexBuf.put(5, 0f);
        m_fBackTexBuf.put(6, (float)m_WindowWidth / MAX_TEXTURE_SIZE);
        m_fBackTexBuf.put(7, 0f);
      /* 设置并载入粒子纹理 */
        gl.glBindTexture(GL10.GL_TEXTURE_2D, m_iTextureBuf.get(0));
        gl.glTexParameterx(GL10.GL_TEXTURE_2D,
                GL10.GL_TEXTURE_MAG_FILTER, GL10.GL_LINEAR);
        gl.glTexParameterx(GL10.GL_TEXTURE_2D,
                GL10.GL_TEXTURE_MIN_FILTER, GL10.GL_LINEAR_MIPMAP_NEAREST);
        gl.glTexParameterx(GL10.GL_TEXTURE_2D,
                GL10.GL_TEXTURE_WRAP_S, GL10.GL_CLAMP_TO_EDGE);
        gl.glTexParameterx(GL10.GL_TEXTURE_2D,
```

```
            GL10.GL_TEXTURE_WRAP_T, GL10.GL_CLAMP_TO_EDGE);
    Utils.generateMipmapsForBoundTexture(texture);
    /* 释放纹理图 */
    texture.recycle();
  /* 读取背景纹理并缩放到 MAX_TEXTURE_SIZE * MAX_TEXTURE_SIZE */
    Bitmap bmp =
      BitmapFactory.decodeResource(m_Context.getResources(),
                                                  R.drawable.back);

    Bitmap background =
itmap.createScaledBitmap(bmp, MAX_TEXTURE_SIZE,
                                         MAX_TEXTURE_SIZE, true);

    /* 设置并载入背景纹理 */
    gl.glBindTexture(GL10.GL_TEXTURE_2D, m_iTextureBuf.get(1));
    gl.glTexParameterx(GL10.GL_TEXTURE_2D,
            GL10.GL_TEXTURE_MAG_FILTER, GL10.GL_NEAREST);
    gl.glTexParameterx(GL10.GL_TEXTURE_2D,
            GL10.GL_TEXTURE_MIN_FILTER, GL10.GL_NEAREST);
    gl.glTexParameterx(GL10.GL_TEXTURE_2D,
            GL10.GL_TEXTURE_WRAP_S, GL10.GL_CLAMP_TO_EDGE);
    gl.glTexParameterx(GL10.GL_TEXTURE_2D,
            GL10.GL_TEXTURE_WRAP_T, GL10.GL_CLAMP_TO_EDGE);
    GLUtils.texImage2D(GL10.GL_TEXTURE_2D, 0, background, 0);
    /* 释放纹理图 */
    bmp.recycle();
    background.recycle();
}
/* 更新背景纹理 */
private void UpdateBackTexture(GL10 gl)
{
    /* 若不是第一次装载纹理,则只从截屏图中更新背景纹理 */
    gl.glEnable(GL10.GL_TEXTURE_2D);
//  Log.v("Update","start");
   int data[] = new int[TEXTURE_SIZE];
    int data_raw[] = new int[SCREEN_SIZE];
    m_bmpBackups[m_iBackup].getPixels(data_raw, 0,
    m_WindowWidth, 0, 0, m_WindowWidth, m_WindowHeight);
   Bitmap background =
Bitmap.createBitmap(MAX_TEXTURE_SIZE,
                    MAX_TEXTURE_SIZE, Config.ARGB_8888);
```

```
        int i;
        int spos = 0, dpos = TEXTURE_SIZE-MAX_TEXTURE_SIZE;
        for (i = 0; i < m_WindowHeight; i++)
        {
            System.arraycopy(data_raw, spos, data, dpos, m_WindowWidth);
            spos += m_WindowWidth;
            dpos -= MAX_TEXTURE_SIZE;
        }
        background.setPixels(data, TEXTURE_SIZE-MAX_TEXTURE_SIZE,
            -MAX_TEXTURE_SIZE, 0, 0, MAX_TEXTURE_SIZE, MAX_TEXTURE_SIZE);
        /* 设置并载入粒子纹理 */
        gl.glBindTexture(GL10.GL_TEXTURE_2D, m_iTextureBuf.get(2));
        gl.glTexParameterx(GL10.GL_TEXTURE_2D,
                GL10.GL_TEXTURE_MAG_FILTER, GL10.GL_LINEAR);
        gl.glTexParameterx(GL10.GL_TEXTURE_2D,
                GL10.GL_TEXTURE_MIN_FILTER, GL10.GL_LINEAR);
        gl.glTexParameterx(GL10.GL_TEXTURE_2D,
                GL10.GL_TEXTURE_WRAP_S, GL10.GL_CLAMP_TO_EDGE);
        gl.glTexParameterx(GL10.GL_TEXTURE_2D,
                GL10.GL_TEXTURE_WRAP_T, GL10.GL_CLAMP_TO_EDGE);
        GLUtils.texImage2D(GL10.GL_TEXTURE_2D, 0, background, 0);
        gl.glDisable(GL10.GL_TEXTURE_2D);
        background.recycle();
        m_bUpdate = false;
    }
```

(15) 图片的绘制

```
public void onDrawFrame(GL10 gl)
    {
        /* 跳过绘制完成的帧,第一帧除外 */
        //if (m_ipHead == m_ipTail && m_ipHead != 0)
        //  return;
        /* 如果当前粒子全部死亡且有截屏信息,则截屏,然后保存在备份图像中 */
        if (m_ipHead == m_ipTail && m_bScreenShot)
        {
            int size = m_WindowWidth * m_WindowHeight;
            /* 32 位缓存 */
            ByteBuffer buf = ByteBuffer.allocateDirect(size * 4);
            buf.order(ByteOrder.nativeOrder());
            /* 读取屏幕数据 */
            gl.glReadPixels(0, 0, m_WindowWidth, m_WindowHeight, GL10.GL_RGBA,
```

```
GL10.GL_UNSIGNED_BYTE, buf);
            /* 将字节数据转为整型数据 */
            int data[] = new int[size];
            buf.asIntBuffer().get(data);
            buf = null;
            /* 将 ABGR 转换成 ARGB */
            int i;
            for (i = 0; i < size; i++)
            {
                data[i] = (data[i] & 0xFF00FF00) | ((data[i] & 0x00FF0000) >> 16) |
                                        ((data[i] & 0x000000FF) << 16);
            }
            //Log.v("IN","IN111");
            /* 修改堆栈和当前备份图像索引 */
            onScreenShot();

            /* 存入当前备份图像中 */
                m_bmpBackups[m_iBackup].setPixels(data, size-m_WindowWidth,
                        -m_WindowWidth, 0, 0, m_WindowWidth,
                                            m_WindowHeight);
                data = null;
                //Log.v("IN","IN222");
                //Log.v("backup",Integer.toString(m_iBackup));
                /* 结束截屏 */
                m_bScreenShot = false;
                /* 跳过该帧 */
                return;
        }
        int i, p;
//      FloatBuffer fbuf = FloatBuffer.allocate(8);
        FloatBuffer fbuf = Utils.newFloatBuffer(8);
        float tmpx, tmpy;

        /* 清空颜色缓存 */
        gl.glClearColor(0, 0, 0, 0);
        /* 清空缓存 */
        gl.glClear(GL10.GL_COLOR_BUFFER_BIT | GL10.GL_DEPTH_BUFFER_BIT);

        /* 将当前矩阵置换为单位矩阵 */
        gl.glLoadIdentity();
```

```
/**
  * Step 1：绘制历史背景
  */
/* 关闭深度测试 */
 gl.glDisable(GL10.GL_DEPTH_TEST);
 /**
  * Step 1(A):若需要更新纹理
  * 则读取当前备份图像作为背景纹理 2
  * 并绘制在屏幕上
  * 然后反色处理
  * 接下来复制屏幕内容到背景纹理 1
  * 再反色处理一次
  * 最后强制埋葬尸体,以防撤销前的图像继续被画在屏幕上
  * 结束当前帧
  */
if (m_bUpdate)
{
  //Log.v("draw","update");
  /** 读取当前备份图像作为背景纹理 2 */
  UpdateBackTexture(gl);
  /** 绘制在屏幕上 */
  /* 设置二维纹理有效 */
  gl.glEnable(GL10.GL_TEXTURE_2D);
  /* 绑定背景纹理为当前纹理 */
  gl.glBindTexture(GL10.GL_TEXTURE_2D, m_iTextureBuf.get(2));
  gl.glColor4f(1, 1, 1, 1);
  /* 开始绘制 */
  gl.glEnableClientState(GL10.GL_VERTEX_ARRAY);
  gl.glEnableClientState(GL10.GL_TEXTURE_COORD_ARRAY);

  /* 保存原有矩阵 */
  gl.glPushMatrix();
  /* 倒置以重合 */
  gl.glRotatef(180, 1, 0, 0);
  /* 绘制 */
  gl.glVertexPointer(2, GL10.GL_FLOAT, 0, m_fBackBuf);
  gl.glTexCoordPointer(2, GL10.GL_FLOAT, 0, m_fBackTexBuf);
  gl.glDrawArrays(GL10.GL_TRIANGLE_FAN, 0, 4);
  /* 恢复原有矩阵 */
  gl.glPopMatrix();
```

```
/* 结束绘制 */
gl.glDisableClientState(GL10.GL_VERTEX_ARRAY);
gl.glDisableClientState(GL10.GL_TEXTURE_COORD_ARRAY);
/** 反色处理 1 */
/* 关闭二维纹理 */
gl.glDisable(GL10.GL_TEXTURE_2D);
/* 开启 Alpha 混色 */
gl.glEnable(GL10.GL_BLEND);
/* 设置混合模式为反色叠加 */
gl.glBlendFunc(GL10.GL_ONE_MINUS_DST_COLOR, GL10.GL_ZERO);
/* 设置当前颜色 */
gl.glColor4f(1, 1, 1, 1);
/* 开始绘制 */
gl.glEnableClientState(GL10.GL_VERTEX_ARRAY);
gl.glVertexPointer(2, GL10.GL_FLOAT, 0, m_fBackBuf);
gl.glDrawArrays(GL10.GL_TRIANGLE_FAN, 0, 4);
/* 结束绘制 */
gl.glDisableClientState(GL10.GL_VERTEX_ARRAY);
/* 关闭 Alpha 混色 */
gl.glDisable(GL10.GL_BLEND);
/** 复制屏幕内容到背景纹理 1 */
/* 开启二维纹理 */
gl.glEnable(GL10.GL_TEXTURE_2D);
/* 绑定背景纹理为当前纹理 */
gl.glBindTexture(GL10.GL_TEXTURE_2D, m_iTextureBuf.get(1));
/* 将尸体粒子固化到背景中 */
gl.glCopyTexSubImage2D(GL10.GL_TEXTURE_2D, 0, 0, 0, 0, 0,
                m_WindowWidth, m_WindowHeight);
/** 反色处理 2 */
/* 关闭二维纹理 */
gl.glDisable(GL10.GL_TEXTURE_2D);
/* 开启 Alpha 混色 */
gl.glEnable(GL10.GL_BLEND);
/* 设置混合模式为反色叠加 */
gl.glBlendFunc(GL10.GL_ONE_MINUS_DST_COLOR, GL10.GL_ZERO);
/* 设置当前颜色 */
gl.glColor4f(1, 1, 1, 1);

/* 开始绘制 */
gl.glEnableClientState(GL10.GL_VERTEX_ARRAY);
```

```
            gl.glVertexPointer(2, GL10.GL_FLOAT, 0, m_fBackBuf);
            gl.glDrawArrays(GL10.GL_TRIANGLE_FAN, 0, 4);
            /* 结束绘制 */
            gl.glDisableClientState(GL10.GL_VERTEX_ARRAY);
            /* 关闭 Alpha 混色 */
            gl.glDisable(GL10.GL_BLEND);
            /** 强制埋葬尸体 */
            m_ipCadaver = m_ipHead;
            // 当前帧绘制结束
            return;
        }
    /**
     * Step 1(B):若不需要更新纹理
     * 则直接绘制背景纹理到屏幕
     */
        /* 设置二维纹理有效 */
        gl.glEnable(GL10.GL_TEXTURE_2D);
        /* 绑定背景纹理为当前纹理 */
        gl.glBindTexture(GL10.GL_TEXTURE_2D, m_iTextureBuf.get(1));
        gl.glColor4f(1, 1, 1, 1);
        /* 开始绘制 */
        gl.glEnableClientState(GL10.GL_VERTEX_ARRAY);
        gl.glEnableClientState(GL10.GL_TEXTURE_COORD_ARRAY);
        /* 绘制 */
        gl.glVertexPointer(2, GL10.GL_FLOAT, 0, m_fBackBuf);
        gl.glTexCoordPointer(2, GL10.GL_FLOAT, 0, m_fBackTexBuf);
        gl.glDrawArrays(GL10.GL_TRIANGLE_FAN, 0, 4);
        /* 结束绘制 */
        gl.glDisableClientState(GL10.GL_VERTEX_ARRAY);
        gl.glDisableClientState(GL10.GL_TEXTURE_COORD_ARRAY);

        /**
        * Step 2: 将刚刚死亡的粒子固化到历史背景中
        */
        /* 开启 Alpha 混色 */
        gl.glEnable(GL10.GL_BLEND);
        /* 设置 Alpha 混色方式 */
        if (m_bInk)
            gl.glBlendFunc(GL10.GL_SRC_ALPHA, GL10.GL_ONE);
        else
```

```
    gl.glBlendFunc(GL10.GL_SRC_COLOR, GL10.GL_ONE_MINUS_SRC_COLOR);
/* 绑定粒子纹理为当前纹理 */
gl.glBindTexture(GL10.GL_TEXTURE_2D, m_iTextureBuf.get(0));
/* 开始绘制 */
gl.glEnableClientState(GL10.GL_VERTEX_ARRAY);
gl.glEnableClientState(GL10.GL_TEXTURE_COORD_ARRAY);

for (i = m_ipCadaver; i < m_ipHead; i++)
{
    p = i % MAX_PARTICLE;
    /* 设置粒子颜色 */
    gl.glColor4f(m_Particles[p].color[0], m_Particles[p].color[1],
                              m_Particles[p].color[2], m_Particles[p].ink);
    /* 简化运算 */
    tmpx = m_Particles[p].size / m_WindowWidth;
    tmpy = m_Particles[p].size / m_WindowHeight;
    /* 设置纹理顶点坐标 */
    fbuf.put(0, m_Particles[p].x + tmpx);
    fbuf.put(1, m_Particles[p].y + tmpy);
    fbuf.put(2, m_Particles[p].x - tmpx);
    fbuf.put(3, m_Particles[p].y + tmpy);
    fbuf.put(4, m_Particles[p].x - tmpx);
    fbuf.put(5, m_Particles[p].y - tmpy);
    fbuf.put(6, m_Particles[p].x + tmpx);
    fbuf.put(7, m_Particles[p].y - tmpx);
    /* 绘制 */
    gl.glVertexPointer(2, GL10.GL_FLOAT, 0, fbuf);
    gl.glTexCoordPointer(2, GL10.GL_FLOAT, 0, m_fTextureBuf);
    gl.glDrawArrays(GL10.GL_TRIANGLE_FAN, 0, 4);
}
/* 结束绘制 */
gl.glDisableClientState(GL10.GL_VERTEX_ARRAY);
gl.glDisableClientState(GL10.GL_TEXTURE_COORD_ARRAY);
if (m_ipHead > m_ipCadaver + DEAD_RESERVE)
{
    /* 绑定背景纹理为当前纹理 */
    gl.glBindTexture(GL10.GL_TEXTURE_2D, m_iTextureBuf.get(1));
    /* 将尸体粒子固化到背景中 */
    gl.glCopyTexSubImage2D(GL10.GL_TEXTURE_2D, 0, 0, 0, 0, 0,
                        m_WindowWidth, m_WindowHeight);
```

```
        /* 埋葬尸体 */
        m_ipCadaver = m_ipHead;
    }
    /**
     ** Step 3: 绘制当前存活的粒子
     */
    /* 绑定粒子纹理为当前纹理 */
    gl.glBindTexture(GL10.GL_TEXTURE_2D, m_iTextureBuf.get(0));
    /* 开始绘制 */
    gl.glEnableClientState(GL10.GL_VERTEX_ARRAY);
    gl.glEnableClientState(GL10.GL_TEXTURE_COORD_ARRAY);
    for (i = m_ipHead; i < m_ipTail; i++)
    {
        p = i % MAX_PARTICLE;
        /* 设置粒子颜色 */
        gl.glColor4f(m_Particles[p].color[0],
            m_Particles[p].color[1], m_Particles[p].color[2], m_Particles[p].ink);
        /* 简化运算 */
        tmpx = m_Particles[p].size / m_WindowWidth;
        tmpy = m_Particles[p].size / m_WindowHeight;
        /* 设置纹理顶点坐标 */
        fbuf.put(0, m_Particles[p].x + tmpx);
        fbuf.put(1, m_Particles[p].y + tmpy);
        fbuf.put(2, m_Particles[p].x - tmpx);
        fbuf.put(3, m_Particles[p].y + tmpy);
        fbuf.put(4, m_Particles[p].x - tmpx);
        fbuf.put(5, m_Particles[p].y - tmpy);
        fbuf.put(6, m_Particles[p].x + tmpx);
        fbuf.put(7, m_Particles[p].y - tmpx);
      /* 绘制 */
        gl.glVertexPointer(2, GL10.GL_FLOAT, 0, fbuf);
        gl.glTexCoordPointer(2, GL10.GL_FLOAT, 0, m_fTextureBuf);
        gl.glDrawArrays(GL10.GL_TRIANGLE_FAN, 0, 4);//考虑优化
    }
    /* 结束绘制 */
    gl.glDisableClientState(GL10.GL_VERTEX_ARRAY);
    gl.glDisableClientState(GL10.GL_TEXTURE_COORD_ARRAY);
    /*
     ** Step 4: 反色绘制背景纹理
     */
```

```
        /* 关闭二维纹理 */
        gl.glDisable(GL10.GL_TEXTURE_2D);

        /* 设置混合模式为反色叠加 */
        gl.glBlendFunc(GL10.GL_ONE_MINUS_DST_COLOR, GL10.GL_ZERO);
        /* 设置当前颜色 */
        gl.glColor4f(1, 1, 1, 1);
        /* 开始绘制 */
        gl.glEnableClientState(GL10.GL_VERTEX_ARRAY);
        gl.glVertexPointer(2, GL10.GL_FLOAT, 0, m_fBackBuf);
        gl.glDrawArrays(GL10.GL_TRIANGLE_FAN, 0, 4);
        /* 结束绘制 */
        gl.glDisableClientState(GL10.GL_VERTEX_ARRAY);
        gl.glDisable(GL10.GL_BLEND);
    }
```

7.6　本章小结

本章通过项目案例学习了 GLSurfaceView 的使用，对 Renderer 进行了深入的了解。对坐标、Canvas 的使用有了进一步的了解。本章项目案例也涉及线程等知识，这些知识点也很重要，争取熟练掌握并运用。

第8章 拼图游戏项目案例

拼图游戏是一款益智娱乐软件,它属于游戏类软件。本章以拼图游戏项目为例介绍游戏类软件项目UI的一般设计方法,以及本软件所涉及的游戏算法、相机使用、图片处理等相关内容。

8.1 预备知识

8.1.1 自定义适配器的应用

很多时候在项目中用到复杂结构的ListView,需要监听其中不同控件的响应,这可以通过自定义适配器来完成。在ListView 3种适配器当中,最受大家青睐的是SimpleAdapter适配器。它的扩展性很强,可以将ListView中每一项都使用自定义布局,插入许多组件;但是SimpleAdapter也有弱点,那就是当ListView中每一项有Button、CheckBox等具有事件的组件,当我们想监听它们时,就必须自定义适配器!

自定义适配器类,是继承自BaseAdapter的类,继承之后会自动重写以下方法:getCount(),getItem(int),getItemId(int),getView(int,View,ViewGroup)。

- getCount()方法:返回我们后台一共有多少数据。
- getItem()与getItemId()两个方法对于Android来讲是没有用的,单纯是为了客户端调用的方便。所以我们可以返回任何对我们有用的值,而不需要顾及Android对它们的使用(因为根本就没有使用)。
- getView方法的工作原理是:ListView针对List中每个item,要求adapter"给我一个视图",那么,一个新的视图将被返回并显示。

如果我们有上亿个项目(item)要显示怎么办?为每个item创建一个新视图?不需要这样做。因为,Android为我们缓存了视图。

Android中有个叫Recycler的构件,图8-1是它的工作原理。

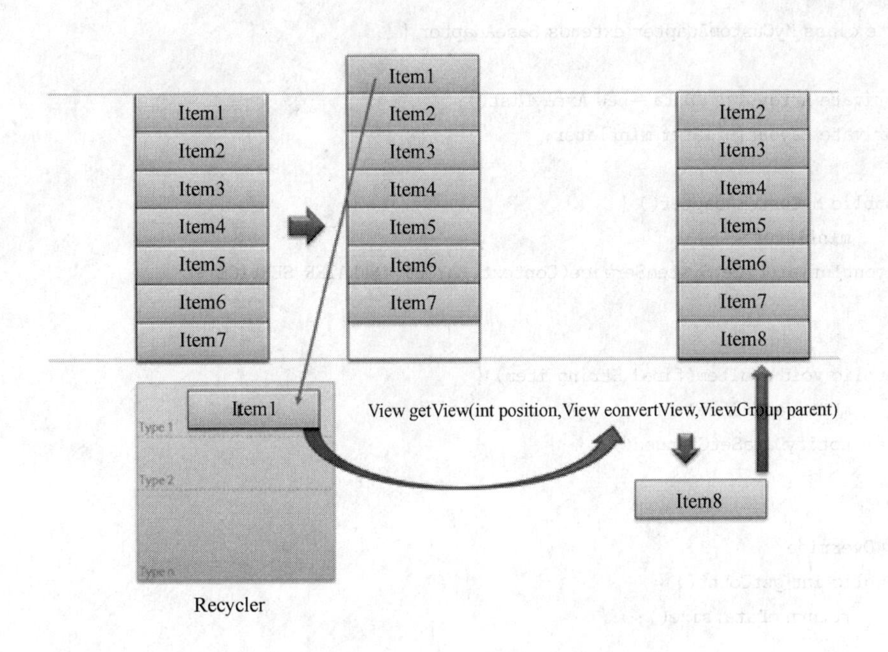

图 8-1 工作原理

(1) 如果我们有 10 亿个项目(item),其中只有可见的项目存在内存中,其他的在Recycler中。

(2) ListView 先请求一个 type1 视图(getView),然后请求其他可见的项目。convertView 在 getView 中是空(null)的。

(3) 当 Item1 滚出屏幕,并且一个新的项目从屏幕低端上来时,ListView 再请求一个 type1 视图。此时,convertView 不是空值了,它的值是 Item1。我们只需设定新的数据然后返回 convertView,不必重新创建一个视图。

请看下面的示例代码,这里在 getView 中使用了 System.out 进行输出:

```
public class MultipleItemsList extends ListActivity {

    private MyCustomAdapter mAdapter;

    @Override
    public void onCreate(Bundle savedInstanceState) {
        super.onCreate(savedInstanceState);
        mAdapter = new MyCustomAdapter();
        for (int i = 0; i < 50; i++) {
            mAdapter.addItem("item " + i);
        }
        setListAdapter(mAdapter);
    }
```

```
private class MyCustomAdapter extends BaseAdapter {

    private ArrayList mData = new ArrayList();
    private LayoutInflater mInflater;

    public MyCustomAdapter() {
        mInflater =
  (LayoutInflater)getSystemService(Context.LAYOUT_INFLATER_SERVICE);
    }

    public void addItem(final String item) {
        mData.add(item);
        notifyDataSetChanged();
    }

    @Override
    public int getCount() {
        return mData.size();
    }

    @Override
    public String getItem(int position) {
        return mData.get(position);
    }

    @Override
    public long getItemId(int position) {
        return position;
    }

    @Override
    public View getView(int position, View convertView, ViewGroup parent) {
        System.out.println("getView " + position + " " + convertView);
        ViewHolder holder = null;
        if (convertView == null) {
            convertView = mInflater.inflate(R.layout.item1, null);
            holder = new ViewHolder();
            holder.textView =
                    (TextView)convertView.findViewById(R.id.text);
            convertView.setTag(holder);
        } else {
            holder = (ViewHolder)convertView.getTag();
        }
        holder.textView.setText(mData.get(position));
        return convertView;
    }
```

```
    }

    public static class ViewHolder {
        public TextView textView;
    }
}
```

初始化几个 item 之后，就不会调用 convertView 的实例化函数，if (convertView==null)里面的部分将不再执行，所以应该在其外设置数据以及相应的监听。

注意：convertView 是 R. layout. item1 中最外面的 layout。

8.1.2　调用系统照相机

因为本案例中，我们调用的是系统的照相机，所以实现方法很简单，只要使用Android封装好的代码直接调用就可以了，实现代码如下：

```
// 调用系统的照相机
Intent intent = new Intent(MediaStore.ACTION_IMAGE_CAPTURE);
intent.putExtra(MediaStore.EXTRA_OUTPUT, fileUri);
startActivityForResult(intent, CAPTURE_IMAGE_ACTIVITY_REQUEST_CODE);
// 得到所照的照片
Bundle bundle = data.getExtras();
Bitmap bitmap = (Bitmap) bundle.get("data");
```

8.1.3　图片处理

Android 为图片切割封装了两个类：ImagePiece 类和 ImageSplitter 类。

(1) ImagePiece 类保存了一个 BitMap 对象和一个表示图片索引的 int 变量。Java 代码如下：

```
import android.graphics.Bitmap;
public class ImagePiece
{
   public int index = 0;
   public Bitmap bitmap = null;
}
```

(2) ImageSplitter 类有一个静态方法 split，传入的参数是要切割的 Bitmap 对象和横、竖向的切割片数。比如传入的是 3,3，则横、竖向都切割成 3 片，最终会将整个图片切割成 3×3=9 片。Java 代码如下：

```
public class ImageSplitter {
public static List<Bitmap> split(Bitmap bitmap, int xPiece, int yPiece) {
// 将一张图片切成 9 张
ArrayList<Bitmap> pieces = new ArrayList<Bitmap>(xPiece * yPiece);
int width = bitmap.getWidth();
// 切割图片的宽
int height = bitmap.getHeight();
// 切割图片的高
```

```
int pieceWidth = width / 3;
    // 切割后每张图片的宽
int pieceHeight = height / 3;
    // 切割后每张图片的高
    ImagePiece piece = new ImagePiece();
    for (int i = 0; i<3; i++) {
      for (int j = 0; j<3; j++){
        piece.index = j+i * xPiece;
        int xValue = j * pieceWidth;
        int yValue = i * pieceHeight;
        Bitmap bit = Bitmap.createBitmap(bitmap, xValue, yValue,
        pieceWidth, pieceHeight);
        pieces.add(bit);
      }
    }
  piece.setBitmap(pieces);
  return pieces;
  }
  }
```

8.1.4 手机屏幕触碰的处理

手机屏幕触碰处理方法是：public Boolean onTouchEvent (MotionEvent event)，其中：

(1) 参数 event。为手机屏幕触摸事件封装类的对象，其中封装了该事件的所有信息，例如，触摸的位置、触摸的类型以及触摸的时间等。该对象会在用户触摸手机屏幕时被创建。

(2) 返回值。该方法的返回值原理与键盘响应事件的原理相同，都是当已经完整地处理了该事件且不希望其他回调方法再次处理时返回 true，否则返回 false。

该方法并不像之前介绍过的方法那样只处理一种事件，一般情况下以下 3 种情况的事件全部由 onTouchEvent 方法处理，只是 3 种情况中的动作值不同。

(1) 手指按下屏幕。当屏幕被按下时，会自动调用该方法来处理事件，此时 MotionEvent. getAction()的值为 MotionEvent. ACTION_DOWN，如果在应用程序中需要处理屏幕被按下的事件，只需重新回调该方法，然后在方法中进行动作的判断即可。

(2) 手指从屏幕离开。当手指离开屏幕时触发的事件，该事件同样需要 onTouchEvent 方法来捕捉，然后在方法中进行动作判断。当 MotionEvent. getAction()的值为 MotionEvent. ACTION_UP 时，表示是屏幕被抬起的事件。

(3) 手指在屏幕中拖动。该方法还负责处理手指在屏幕上滑动的事件，同样是调用 MotionEvent. getAction()方法来判断动作值是否为 MotionEvent. ACTION_MOVE，再进行处理。

8.2　需求分析

(1) 选取本地图片，将图片切割成9份，打乱顺序，完成拼图游戏。

(2) 调用本地相机，拍完照以后得到所拍照片，将照片切割成9份，打乱顺序，完成拼图游戏。

8.3　功能分析

用户可以在本地选取图片，然后将图片打乱顺序，实现拼图游戏，也可以通过相机拍照片，将照片打乱顺序，实现拼图游戏。

8.4　设　　计

拼图游戏软件的界面设计与功能描述如表8-1所示。

表8-1　拼图游戏软件的界面设计与功能描述

界面	功能描述
主界面	1. 信息显示 开始游戏 2. 功能按钮 开始游戏按钮
选择拼图类型	1. 信息显示 拼图类型 2. 功能按钮 本地图片按钮 自己制作图片按钮
选择本地图片	1. 信息显示 本地可供选择的图片 2. 功能按钮 无
选择自己制作新图片	1. 信息显示 系统照相机 2. 功能按钮 拍照按钮 取消按钮 重拍按钮
拼图界面	1. 信息显示 需要进行拼图的图片 2. 功能按钮 开始按钮

8.4.1 UI设计

(1) 单击“开始”按钮进行游戏，如图 8-2 所示。

(2) 单击“选取本地拼图” 进入本地图片的选择；单击“制作新的拼图”进入系统相机，如图 8-3 所示。

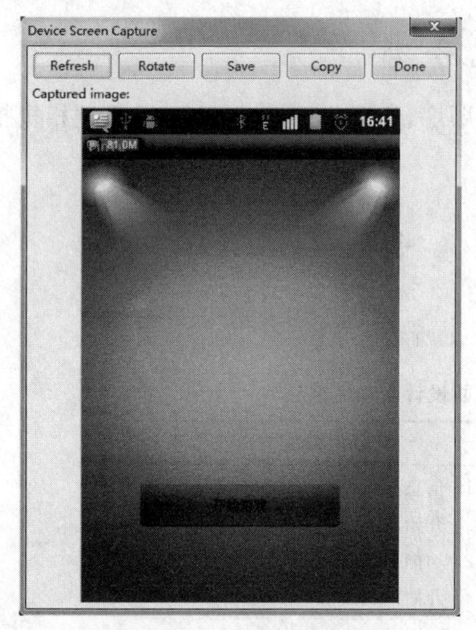

图 8-2 “开始”界面

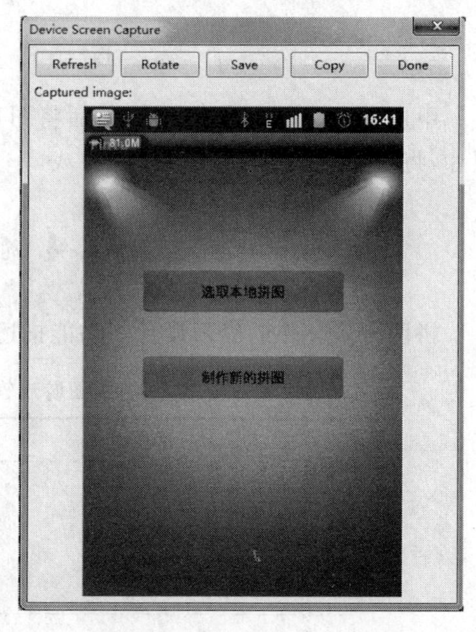

图 8-3 “选择模式”界面

(3) 单击大图片进入“拼图”界面，如图 8-4 所示。

图 8-4 “拼图”界面

(4) 单击“开始”按钮打乱图片,如图 8-5 所示。

(a)

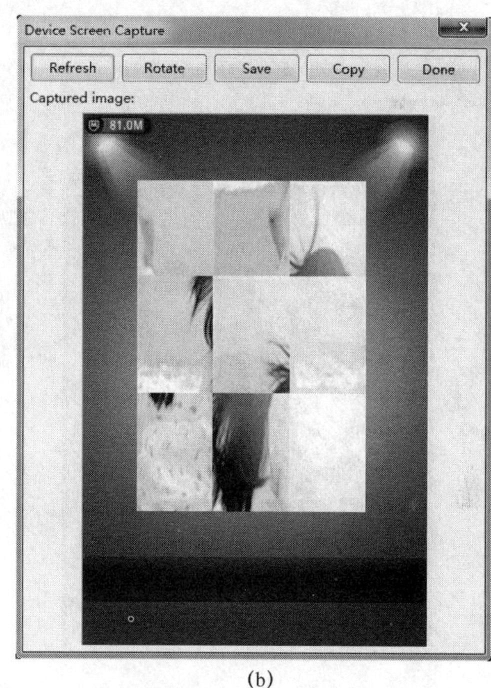

(b)

图 8-5　“开始游戏”界面 1

(5) 调用系统相机界面,如图 8-6 所示。

(6) “拍照”界面如图 8-7 所示。

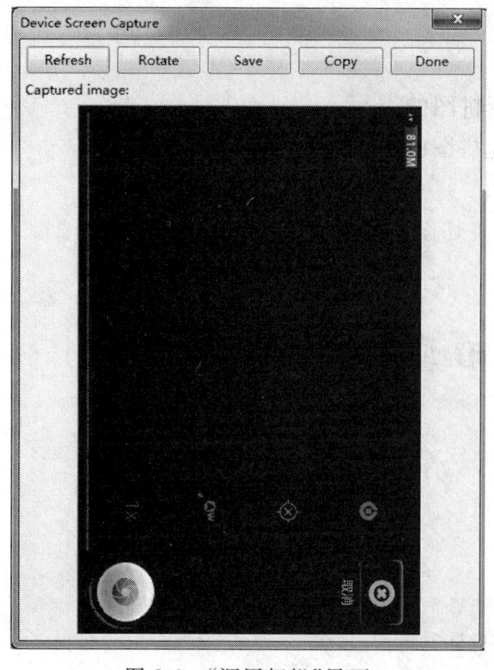

图 8-6　“调用相机”界面

图 8-7　“拍照”界面

(7) 单击“开始”按钮打乱已经拍照好的图片,如图 8-8 所示。

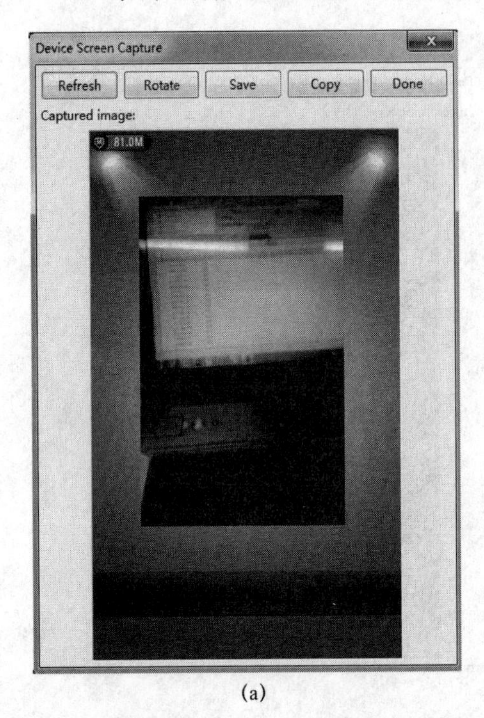

(a)

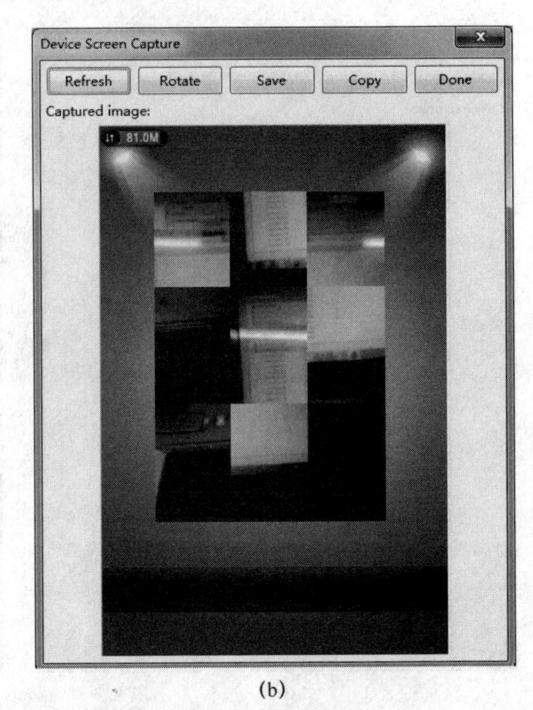

(b)

图 8-8 “开始游戏”界面 2

8.4.2 类设计

软件中用到的类如下:

(1) MyActivity. java。主页面的类。

(2) Choose. java。选择本地拼图/制作新拼图的类。

(3) Android_GalleryActivity. java。本地拼图的类。

(4) MyCamera. java。制作新拼图。

(5) ImagePiece. java,ImageSplitter. java。切图的类。

(6) Puzzle. java。拼图的类。

8.5 编程实现

8.5.1 XML 布局

(1) 主页面布局

activity_main. xml

```
<RelativeLayout xmlns:android="http://schemas.android.com/apk/res/android"
    xmlns:tools="http://schemas.android.com/tools"
    android:layout_width="match_parent"
```

```
android:layout_height = "match_parent"
android:background = "@drawable/loginbg">
<Button
    android:id = "@ + id/button1"
    android:layout_width = "wrap_content"
    android:layout_height = "wrap_content"
    android:layout_alignParentBottom = "true"
    android:layout_centerHorizontal = "true"
    android:layout_marginBottom = "74dp"
    android:onClick = "start"
    android:text = "开始游戏"
    android:background = "@drawable/arowss"/>

</RelativeLayout>
```

(2) 选择拼图类型界面布局

choose.xml

```
<RelativeLayout xmlns:android = "http://schemas.android.com/apk/res/android"
    android:layout_width = "fill_parent"
    android:layout_height = "fill_parent"
    android:background = "@drawable/loginbg">

    <Button
        android:id = "@ + id/button1"
        android:layout_width = "wrap_content"
        android:layout_height = "wrap_content"
        android:layout_alignParentTop = "true"
        android:layout_centerHorizontal = "true"
        android:layout_marginTop = "110dp"
        android:text = "选取本地拼图"
        android:onClick = "start"
        android:background = "@drawable/arowss"/>

    <Button
        android:id = "@ + id/button2"
        android:layout_width = "wrap_content"
        android:layout_height = "wrap_content"
        android:layout_alignRight = "@ + id/button1"
        android:layout_centerVertical = "true"
        android:text = "制作新的拼图"
        android:onClick = "choose"
        android:background = "@drawable/arowss"/>
```

```
</RelativeLayout>
```

(3) 选择本地拼图界面布局

kuangjia.xml

```
<?xml version="1.0" encoding="utf-8"?>
<LinearLayout xmlns:android="http://schemas.android.com/apk/res/android"
    android:orientation="vertical" android:layout_width="fill_parent"
    android:layout_height="fill_parent">
    <Gallery
        android:id="@+id/gallery1"
        android:layout_width="match_parent"
        android:layout_height="wrap_content"
        android:spacing="20px"></Gallery>
    <ImageView
        android:id="@+id/imagepreview1"
        android:layout_height="fill_parent"
        android:layout_width="fill_parent"
        android:scaleType="fitCenter"
        android:layout_marginTop="10px"
        android:clickable="true"
        android:onClick="onClick"/>
    </LinearLayout>
```

(4) 为 Gallery 做模板的布局

image.xml

```
<?xml version="1.0" encoding="utf-8"?>
<ImageView
   xmlns:android="http://schemas.android.com/apk/res/android"
   android:orientation="vertical"
   android:layout_width="80px"
   android:layout_height="120px"
   android:id="@+id/ImageView2"
   android:background="@android:drawable/gallery_thumb">
</ImageView>
```

(5) 拼图界面布局

main.xml

```
<?xml version="1.0" encoding="utf-8"?>
<AbsoluteLayout xmlns:android="http://schemas.android.com/apk/res/android"
    android:layout_width="fill_parent"
    android:layout_height="fill_parent"
    android:background="@drawable/loginbg"
    android:orientation="vertical">
    <ImageView
```

```
    android:id="@+id/ImageView01"
    android:layout_width="wrap_content"
    android:layout_height="wrap_content"
    android:layout_x="50dp"
    android:layout_y="60dp"
    android:padding="0dp"
    android:scaleType="fitXY"
    android:background="@drawable/a1"/>
<ImageView
    android:id="@+id/ImageView02"
    android:layout_width="wrap_content"
    android:layout_height="wrap_content"
    android:layout_x="120dp"
    android:layout_y="60dp"
    android:padding="0dp"
    android:scaleType="fitXY"
    android:background="@drawable/a1"/>
<ImageView
    android:id="@+id/ImageView03"
    android:layout_width="wrap_content"
    android:layout_height="wrap_content"
    android:layout_x="191dp"
    android:layout_y="60dp"
    android:padding="0dp"
    android:scaleType="fitXY"
    android:background="@drawable/a1" />
<ImageView
    android:id="@+id/ImageView04"
    android:layout_width="wrap_content"
    android:layout_height="wrap_content"
    android:layout_x="50dp"
    android:layout_y="146dp"
    android:padding="0dp"
    android:scaleType="fitXY"
    android:background="@drawable/a1"/>
<ImageView
    android:id="@+id/ImageView05"
    android:layout_width="wrap_content"
    android:layout_height="wrap_content"
    android:layout_x="120dp"
    android:layout_y="146dp"
    android:padding="0dp"
```

```
        android:scaleType="fitXY"
        android:background="@drawable/a1"/>
    <ImageView
        android:id="@+id/ImageView06"
        android:layout_width="wrap_content"
        android:layout_height="wrap_content"
        android:layout_x="191dp"
        android:layout_y="146dp"
        android:padding="0dp"
        android:scaleType="fitXY"
        android:background="@drawable/a1"/>
    <ImageView
        android:id="@+id/ImageView07"
        android:layout_width="wrap_content"
        android:layout_height="wrap_content"
        android:layout_x="50dp"
        android:layout_y="252dp"
        android:padding="0dp"
        android:scaleType="fitXY"
        android:background="@drawable/a1"/>
    <ImageView
        android:id="@+id/ImageView08"
        android:layout_width="wrap_content"
        android:layout_height="wrap_content"
        android:layout_x="120dp"
        android:layout_y="252dp"
        android:padding="0dp"
        android:scaleType="fitXY"
        android:background="@drawable/a1"/>
    <ImageView
        android:id="@+id/ImageView09"
        android:layout_width="wrap_content"
        android:layout_height="wrap_content"
        android:layout_x="191dp"
        android:layout_y="252dp"
        android:padding="0dp"
        android:scaleType="fitXY"
        android:background="@drawable/a1"/>
    <Button
        android:id="@+id/button01"
        android:layout_width="fill_parent"
        android:layout_height="wrap_content"
```

```
        android:layout_y="400dp"
        android:onClick="statr"
        android:text="开始"
        android:background="@drawable/arowss" />
</AbsoluteLayout>
```

8.5.2 代码实现

(1) 通过截图可以看出,刚启动游戏的时候进入主页面,单击"开始游戏"按钮进行游戏。

第一步,编写首页面,编码实现如下:

MyActivity

```
public class MyActivity extends Activity{
 protected void onCreate(Bundle savedInstanceState) {
    super.onCreate(savedInstanceState);
      setContentView(R.layout.activity_main);
 }

 /**
  * 单击"开始"按钮开始游戏
  **/
 public void start(View view){
          //跳转到Choose页面,选择拼图的类型
       Intent intent = new Intent(this,Choose.class);
       startActivity(intent);

    }
  }
```

第二步,当单击"开始游戏"按钮时,页面会跳转到选择拼图类型的界面,可以根据自己的喜好选择是用本地图片进行拼图还是利用照相机自己制作图片进行拼图,编码实现如下:

Choose

```
public class Choose extends Activity{
  protected void onCreate(Bundle savedInstanceState) {
       super.onCreate(savedInstanceState);
       setContentView(R.layout.choose); //加载布局
  }

  /**
    *  "选取本地拼图"按钮单击事件
    **/
  public void start(View view){
```

```
        // 单击"选取本地拼图"按钮,进入选取本地图片界面
    Intent intent = new Intent(this,Android_GalleryActivity.class);
    startActivity(intent);
  }

  /**
   * "制作新的拼图"按钮单击事件
   **/
  public void choose(View view){
        // 单击"制作新的图片"按钮进入调用系统照相机界面
    Intent intent = new Intent(this,MyCamera.class);
    startActivity(intent);
  }
}
```

第三步,当单击"选择本地图片"按钮时,页面会跳转到选择本地图片的界面。

首先,我们要为本地加载一些图片,将图片放到工程下的 Drawable 目录里,在代码中将 drawable 里的图片加载到一个数组里,代码实现如下:

Android_GalleryActivity

```
public class Android_GalleryActivity extends Activity {
  int t;
  private ImageView imageView = null;
  private Gallery mGallery    = null;
  // 本地图片资源
  int[] images = { R.drawable.image_1, R.drawable.image_2,
          R.drawable.image_3, R.drawable.image_4, R.drawable.image_5,
          R.drawable.image_6, R.drawable.image_7, R.drawable.image_8,
          R.drawable.image_9, R.drawable.image_10, R.drawable.image_11,
          R.drawable.image_12, R.drawable.image_13, R.drawable.image_14,
          R.drawable.image_15, R.drawable.image_16, R.drawable.image_17,
          R.drawable.image_18, R.drawable.image_19, R.drawable.image_20,
          R.drawable.image_21, R.drawable.image_22, R.drawable.image_23 };
  …
```

然后,在选择本地图片的时候我们采取的是以画廊的形式展示图片,所以我们要在布局里加入 Gallery 布局,代码实现如下:

```
public void onCreate(Bundle savedInstanceState) {
    super.onCreate(savedInstanceState);
    setContentView(R.layout.kuangjia);
    // 初始化 Gallery 控件
    mGallery = (Gallery) this.findViewById(R.id.gallery1);
      // 滑动画廊时,在画廊底下与之相配的大图的 ImageView
    imageView = (ImageView) this.findViewById(R.id.imagepreview1);
```

既然我们已经构建了Gallery,那么接下来我们就要为Gallery适配图片,在本游戏里我们采用的是自定义适配器,其代码实现如下:

```
/**
  * 自定义适配器
  **/
 class MyAdapter extends BaseAdapter {

     // getCount 必须重写
     public int getCount() {
         return images.length;
     }

     // 一般不重写
     public Object getItem(int position) {
         return images[position];
     }

     // 一般不重写
     public long getItemId(int position) {
         return position * 100 + 1;
     }

     // 必须重写,用于创建每一项时调用
     public View getView(int position, View convertView, ViewGroup parent) {
         // 使用布局文件 R.layout.image 为画廊里用于盛放本地图片的 ImageView
         ImageView image = (ImageView) Android_GalleryActivity.this
                 .getLayoutInflater().inflate(R.layout.image, parent, false);
         // 将选中的图片适配到 ImageView 上
         image.setImageResource(images[position]);
         return image;
     }
 }
```

适配器写好后,我们就要为Gallery加载适配器了,并为Gallery注册监听,当选择了一张图片以后,就要将选中的图片适配到相应的ImageView上,以便以后对其进行打乱顺序,实现拼图的目的代码实现如下:

```
// 初始化自定义适配器
     MyAdapter mAdapter = new MyAdapter();
     mGallery.setAdapter(mAdapter);
     // 为 gallery 注册事件,当选中项发生变化后执行
     mGallery.setOnItemSelectedListener(new OnItemSelectedListener() {
         public void onItemSelected(AdapterView<? > arg0, View arg1,
```

```
                int position, long id) {
                //  单击图片的位置
                t = position;
                //  将单击的图片适配到 ImageView 上
                imageView.setImageResource(images[position]);
            }
            public void onNothingSelected(AdapterView<? > arg0) {

            }
        });
    }
```

第四步，当我们选择了一张图片以后，就要进行对图片的切割了，在切割我们所选择的图片之前，我们要先看看切割图片有什么方法。Android 为图片切割封装了两个类：分别是 ImagePiece 类 和 ImageSplitter 类。ImagePiece 类保存了一个 BitMap 对象和一个表示图片索引的 int 变量。代码实现如下：

```
public class ImagePiece extends Application{
  public int index = 0;
    public static ArrayList <Bitmap> bitmap = null;
  public ArrayList<Bitmap> getBitmap() {
      return bitmap;
  }
  public void setBitmap(ArrayList<Bitmap> bitmap) {
      this.bitmap = bitmap;
  }
}
```

Android 为切割图片封装的第二个类为 ImageSplitter 类，此类有一个静态方法 split，传入的参数是要切割的 Bitmap 对象和横向和竖向的切割片数。代码实现如下：

```
public class ImageSplitter {
  public static List<Bitmap> split(Bitmap bitmap, int xPiece, int yPiece) {
      // 将一张图片切成 9 张
      ArrayList<Bitmap> pieces = new ArrayList<Bitmap>(xPiece * yPiece);
      int width = bitmap.getWidth(); // 切割图片的宽
      int height = bitmap.getHeight(); // 切割图片的高
      int pieceWidth = width / 3; // 切割后每张图片的宽
      int pieceHeight = height / 3; // 切割后每张图片的高
      ImagePiece piece = new ImagePiece();
      for (int i = 0; i<3; i++) {
          for (int j = 0; j<3; j++) {
              piece.index = j + i * xPiece;
              int xValue = j * pieceWidth;
              int yValue = i * pieceHeight;
```

```
            Bitmap bit = Bitmap.createBitmap(bitmap, xValue, yValue,
                    pieceWidth, pieceHeight);
            pieces.add(bit);
        }
    }
    // ImagePiece imagePiece = new ImagePiece();
    piece.setBitmap(pieces);
    return pieces;
  }
}
```

第五步,已经建好了 Android 为切割图片封装好的两个类了,下一步我们就要调用这两个类里的方法,以实现对自己所选择图片的切割,代码实现很简单,只要在 Android_GalleryActivity 类里,所单击的 ImageView 的监听里加上如下代码即可,代码如下:

```
/**
  * 大图单击事件:
  * BitmapFactory.decodeResource(?,?,?)
  * 参数 1:包含要加载的位图资源文件的对象一般写成 getResources。
  * 参数 2:需要加载的位图的资源 ID。
  * 参数 3,4:对待加载的位图是否需要完整显示,若只需要部分,可以在这里定制。
  **/
    public void onClick(View view){ // 大图 ImageView 的监听
        int xPiece = 3;   // 横向切割 3 份
        int yPiece = 3;   // 纵向切割 3 份
        // 调用切割图片的方法
    ImageSplitter.split(BitmapFactory.decodeResource(getResources(),
                images[t]),xPiece,yPiece);
        Intent intent = new Intent(this,Puzzle.class);
        startActivity(intent);
```

第六步,切割图片的功能已经实现,下面我们来完成本游戏的核心功能——拼图。首先我们为拼图作些准备工作,具体代码实现如下:

Puzzle

```
public class Puzzle extends Activity {
  ImagePiece piece;
    // 盛放打乱顺序的图片的数组
  Bitmap[] queue = new Bitmap[9];
  // 盛放没打乱顺序的图片的数组
  Bitmap[] queue1 = new Bitmap[9];
  // ImageView 图片显示区域,按九宫格分为 9 个区域
  private ImageView iv_1, iv_2, iv_3, iv_4, iv_5, iv_6, iv_7, iv_8, iv_9;
  // 图片的位置参数对象,
  // 通过设置 LayoutParams 方法的 x,y 坐标,并传给图片,可以改变图片的位置,如:
```

```
// setLayoutParams(paramsIv_1)
private AbsoluteLayout.LayoutParams paramsIv_1, paramsIv_2, paramsIv_3,
        paramsIv_4, paramsIv_5, paramsIv_6, paramsIv_7, paramsIv_8,
        paramsIv_9;
//记录图片的长宽,与左上角一点的坐标
private int picWidth, picHeight, picX1, picY1, picX2, picY2, picX3, picY3,
        picX4, picY4, picX5, picY5, picX6, picY6, picX7, picY7, picX8,
        picY8, picX9, picY9;
//在 onTouchEvent 方法中设置,标记是否击中图片
private boolean clickPic1, clickPic2, clickPic3, clickPic4, clickPic5,
        clickPic6, clickPic7, clickPic8, clickPic9;
//图片锁,每次只能对一个图片进行操作
//flag 为真时表明已经 ACTION_DOWN 一张图片
private boolean flag = false;
```

通过前面的操作,我们已经把切割好的小图片放到了 ImagePiece 类里的 bitmap 集合里,现在我们要在类 Puzzle 类里得到这些切割好的图片,并且将它们分别放到两个数组里,一个数组存放正确顺序的图片,另一个数组里存放打乱顺序的图片,以便最后判断拼图是否拼的正确,代码实现如下:

```
protected void onCreate(Bundle savedInstanceState) {
    super.onCreate(savedInstanceState);
    //去掉标题全屏显示
    requestWindowFeature(Window.FEATURE_NO_TITLE);
    getWindow().setFlags(WindowManager.LayoutParams.FLAG_FULLSCREEN,
            WindowManager.LayoutParams.FLAG_FULLSCREEN);
    //加载 layout
    setContentView(R.layout.main);
    //初始化控件和放置图片
    ArrayList<Bitmap> list = ImagePiece.bitmap;
    for (int i = 0; i<9; i++) {
        queue[i] = list.get(i);     //将对此数组里的图片进行顺序的打乱
        queue1[i] = list.get(i);   //盛放图片正确位置的数组
    }
    findImageView(queue);
}


Bitmap temp; //中间容器,用于图片位置的交换
```

既然图片已经切割好,并且盛放每张小图的 ImageView 也已经准备好了,那么我们现在要做的就是将每张小的图片分别适配到相应的 ImageView 上,编写 findImageView ()方法,具体代码实现如下:

```
/**
 * 将每幅小图加载到各个 ImageView 上
```

```
 **/
@SuppressWarnings("deprecation")
public void findImageView(Bitmap[] queue) {
    // 每幅分图的大小
    picHeight = 106;
    picWidth = 70;
    // 初始化控件
    iv_1 = (ImageView) this.findViewById(R.id.ImageView01);

    // 将切割的第一幅图片适配到相应的 ImageView 上
    iv_1.setImageBitmap(queue[0]);

    // ScaleType.FIT_XY 把图片按比例_扩大/缩小到 View 的大小显示,
    iv_1.setScaleType(ScaleType.FIT_XY);

    // 通过 Drawable 对象的 getLayoutParams()方法取得,可得到图片的位置信息
    paramsIv_1 = (LayoutParams) iv_1.getLayoutParams();
    picX1 = paramsIv_1.x;
    picY1 = paramsIv_1.y;

    iv_2 = (ImageView) this.findViewById(R.id.ImageView02);
    iv_2.setImageBitmap(queue[1]);
    iv_2.setScaleType(ScaleType.FIT_XY);
    paramsIv_2 = (LayoutParams) iv_2.getLayoutParams();
    picX2 = paramsIv_2.x;
    picY2 = paramsIv_2.y;

    iv_3 = (ImageView) this.findViewById(R.id.ImageView03);
    iv_3.setImageBitmap(queue[2]);
    iv_3.setScaleType(ScaleType.FIT_XY);
    paramsIv_3 = (LayoutParams) iv_3.getLayoutParams();
    picX3 = paramsIv_3.x;
    picY3 = paramsIv_3.y;

    iv_4 = (ImageView) this.findViewById(R.id.ImageView04);
    iv_4.setImageBitmap(queue[3]);
    iv_4.setScaleType(ScaleType.FIT_XY);
    paramsIv_4 = (LayoutParams) iv_4.getLayoutParams();
    picX4 = paramsIv_4.x;
    picY4 = paramsIv_4.y;

    iv_5 = (ImageView) this.findViewById(R.id.ImageView05);
```

```
        iv_5.setImageBitmap(queue[4]);
        iv_5.setScaleType(ScaleType.FIT_XY);
        paramsIv_5 = (LayoutParams) iv_5.getLayoutParams();
        picX5 = paramsIv_5.x;
        picY5 = paramsIv_5.y;

        iv_6 = (ImageView) this.findViewById(R.id.ImageView06);
        iv_6.setImageBitmap(queue[5]);
        iv_6.setScaleType(ScaleType.FIT_XY);
        paramsIv_6 = (LayoutParams) iv_6.getLayoutParams();
        picX6 = paramsIv_6.x;
        picY6 = paramsIv_6.y;

        iv_7 = (ImageView) this.findViewById(R.id.ImageView07);
        iv_7.setImageBitmap(queue[6]);
        iv_7.setScaleType(ScaleType.FIT_XY);
        paramsIv_7 = (LayoutParams) iv_7.getLayoutParams();
        picX7 = paramsIv_7.x;
        picY7 = paramsIv_7.y;

        iv_8 = (ImageView) this.findViewById(R.id.ImageView08);
        iv_8.setImageBitmap(queue[7]);
        iv_8.setScaleType(ScaleType.FIT_XY);
        paramsIv_8 = (LayoutParams) iv_8.getLayoutParams();
        picX8 = paramsIv_8.x;
        picY8 = paramsIv_8.y;

        iv_9 = (ImageView) this.findViewById(R.id.ImageView09);
        iv_9.setImageBitmap(queue[8]);
        iv_9.setScaleType(ScaleType.FIT_XY);
        paramsIv_9 = (LayoutParams) iv_9.getLayoutParams();
        picX9 = paramsIv_9.x;
        picY9 = paramsIv_9.y;
}
```

图片切割成功，单击“开始游戏”按钮实现图片的打乱，我们通过对随机数的应用，将图片进行打乱，具体实现如下：

```
// 打乱图片
public static Bitmap[] shuffle(Bitmap[] intArray) {
        Random ran = new Random(); // 随机数
        int num;
        for (int i = 0; i<9; i++) {
```

```
            for (int j = 0; j<9; j++) {
                num = ran.nextInt(9); // 生成一个随机的 int 值,该值介于[0~9)的区间
                swap(intArray, j, num);
            }
        }
        return intArray;
    }

    //用于交换 intArray 中下标分别为 index1 和 index2 的值
    public static void swap(Bitmap[] intArray, int index1, int index2) {
        Bitmap temp = null;
        temp = intArray[index1];
        intArray[index1] = intArray[index2];
        intArray[index2] = temp;
    }

  // 单击"开始游戏"按钮打乱图片顺序,重新布局
    public void statr(View view) {
        queue = shuffle(queue);
        findImageView(queue);
    }
```

每张小图片都已经是配好了,并且顺序已经打乱,我们就可以进行拼图游戏了,下面我们就要完成手机屏幕单击事件的功能,重写 onTouchEvent 方法,代码实现如下:

```
  /**
      * 手机屏幕单击事件
      **/
    public boolean onTouchEvent(MotionEvent event) {
        float x = event.getX();  //  获取屏幕触摸点的横坐标
        float y = event.getY();  //  获取屏幕触摸点的纵坐标

        switch (event.getAction()) {
        case MotionEvent.ACTION_DOWN:  // 手指按下
            if (flag) {
                break;
            } else {
                flag = true;
                // if 语句 判断当鼠标点下时,是否击中图片
                if (x >= picX1 && x <= picX1 + picWidth && y >= picY1
                            && y <= picY1 + picHeight) {
                    clickPic1 = true; // 单击图片
                } else {
```

```
        clickPic1 = false;
    }

    if (x >= picX2 && x <= picX2 + picWidth && y >= picY2
            && y <= picY2 + picHeight) {
        clickPic2 = true;
    } else {
        clickPic2 = false;
    }

    if (x >= picX3 && x <= picX3 + picWidth && y >= picY3
            && y <= picY3 + picHeight) {
        clickPic3 = true;
    } else {
        clickPic3 = false;
    }

    if (x >= picX4 && x <= picX4 + picWidth && y >= picY4
            && y <= picY4 + picHeight) {
        clickPic4 = true;
    } else {
        clickPic4 = false;
    }

    if (x >= picX5 && x <= picX5 + picWidth && y >= picY5
            && y <= picY5 + picHeight) {
        clickPic5 = true;
    } else {
        clickPic5 = false;
    }

    if (x >= picX6 && x <= picX6 + picWidth && y >= picY6
            && y <= picY6 + picHeight) {
        clickPic6 = true;
    } else {
        clickPic6 = false;
    }

    if (x >= picX7 && x <= picX7 + picWidth && y >= picY7
            && y <= picY7 + picHeight) {
        clickPic7 = true;
    } else {
```

```
            clickPic7 = false;
        }

        if (x >= picX8 && x <= picX8 + picWidth && y >= picY8
                && y <= picY8 + picHeight) {
            clickPic8 = true;
        } else {
            clickPic8 = false;
        }

        if (x >= picX9 && x <= picX9 + picWidth && y >= picY9
                && y <= picY9 + picHeight) {
            clickPic9 = true;
        } else {
            clickPic9 = false;
        }
        break;
    }

case MotionEvent.ACTION_UP:  // 手指抬起
    // 保持 ImageView 的位置不变
    paramsIv_1.x = picX1;
    paramsIv_1.y = picY1;
    iv_1.setLayoutParams(paramsIv_1);

    paramsIv_2.x = picX2;
    paramsIv_2.y = picY2;
    iv_2.setLayoutParams(paramsIv_2);

    paramsIv_3.x = picX3;
    paramsIv_3.y = picY3;
    iv_3.setLayoutParams(paramsIv_3);

    paramsIv_4.x = picX4;
    paramsIv_4.y = picY4;
    iv_4.setLayoutParams(paramsIv_4);

    paramsIv_5.x = picX5;
    paramsIv_5.y = picY5;
    iv_5.setLayoutParams(paramsIv_5);

    paramsIv_6.x = picX6;
```

```
            paramsIv_6.y = picY6;
            iv_6.setLayoutParams(paramsIv_6);

            paramsIv_7.x = picX7;
            paramsIv_7.y = picY7;
            iv_7.setLayoutParams(paramsIv_7);

            paramsIv_8.x = picX8;
            paramsIv_8.y = picY8;
            iv_8.setLayoutParams(paramsIv_8);

            paramsIv_9.x = picX9;
            paramsIv_9.y = picY9;
            iv_9.setLayoutParams(paramsIv_9);

            flag = false;

            if (clickPic1) {   // 第 1 张图片被击中
                // picWidth 与 picHeight 分别除以 2,定位击中点为图片的中心
/* 当屏幕触摸点的横坐标,在图片左上角的横坐标与图片右上角的横坐标之间时,
纵坐标同理,满足以上两个条件证明图片被点中 */
                if (x >= picX2 && x <= picX2 + picWidth && y >= picY2
                        && y <= picY2 + picHeight) {
                    // 交换全局 int 数组 queue 的元素顺序,达到交换图片位置的目的
                    temp = queue[0];
                    queue[0] = queue[1];
                    queue[1] = temp;
                    // 显示交换顺序后的图片
                    findImageView(queue);
                } else if (x >= picX3 && x <= picX3 + picWidth && y >= picY3
                        && y <= picY3 + picHeight) {
                    temp = queue[0];
                    queue[0] = queue[2];
                    queue[2] = temp;
                    findImageView(queue);
                } else if (x >= picX4 && x <= picX4 + picWidth && y >= picY4
                        && y <= picY4 + picHeight) {
                    temp = queue[0];
                    queue[0] = queue[3];
                    queue[3] = temp;
                    findImageView(queue);
```

```
    } else if (x >= picX5 && x <= picX5 + picWidth && y >= picY5
            && y <= picY5 + picHeight) {
        temp = queue[0];
        queue[0] = queue[4];
        queue[4] = temp;
        findImageView(queue);
    } else if (x >= picX6 && x <= picX6 + picWidth && y >= picY6
            && y <= picY6 + picHeight) {
        temp = queue[0];
        queue[0] = queue[5];
        queue[5] = temp;
        findImageView(queue);
    } else if (x >= picX7 && x <= picX7 + picWidth && y >= picY7
            && y <= picY7 + picHeight) {
        temp = queue[0];
        queue[0] = queue[6];
        queue[6] = temp;
        findImageView(queue);
    } else if (x >= picX8 && x <= picX8 + picWidth && y >= picY8
            && y <= picY8 + picHeight) {
        temp = queue[0];
        queue[0] = queue[7];
        queue[7] = temp;
        findImageView(queue);
    } else if (x >= picX9 && x <= picX9 + picWidth && y >= picY9
            && y <= picY9 + picHeight) {
        temp = queue[0];
        queue[0] = queue[8];
        queue[8] = temp;
        findImageView(queue);
    } else {
    }
}

if (clickPic2) {   // 第 2 张图片被击中
    // picWidth 与 picHeight 分别除以 2,定位击中点为图片的中心
    if (x >= picX1 && x <= picX1 + picWidth && y >= picY1
            && y <= picY1 + picHeight) {
        temp = queue[0];
        queue[0] = queue[1];
        queue[1] = temp;
        findImageView(queue);
```

```
} else if (x >= picX3 && x <= picX3 + picWidth && y >= picY3
        && y <= picY3 + picHeight) {
    temp = queue[1];
    queue[1] = queue[2];
    queue[2] = temp;
    findImageView(queue);
} else if (x >= picX4 && x <= picX4 + picWidth && y >= picY4
        && y <= picY4 + picHeight) {
    temp = queue[1];
    queue[1] = queue[3];
    queue[3] = temp;
    findImageView(queue);
} else if (x >= picX5 && x <= picX5 + picWidth && y >= picY5
        && y <= picY5 + picHeight) {
    temp = queue[1];
    queue[1] = queue[4];
    queue[4] = temp;
    findImageView(queue);
} else if (x >= picX6 && x <= picX6 + picWidth && y >= picY6
        && y <= picY6 + picHeight) {
    temp = queue[1];
    queue[1] = queue[5];
    queue[5] = temp;
    findImageView(queue);
} else if (x >= picX7 && x <= picX7 + picWidth && y >= picY7
        && y <= picY7 + picHeight) {
    temp = queue[1];
    queue[1] = queue[6];
    queue[6] = temp;
    findImageView(queue);
} else if (x >= picX8 && x <= picX8 + picWidth && y >= picY8
        && y <= picY8 + picHeight) {
    temp = queue[1];
    queue[1] = queue[7];
    queue[7] = temp;
    findImageView(queue);
} else if (x >= picX9 && x <= picX9 + picWidth && y >= picY9
        && y <= picY9 + picHeight) {
    temp = queue[1];
    queue[1] = queue[8];
    queue[8] = temp;
    findImageView(queue);
```

```
    } else {
    }
}

    if (clickPic3) {   // 第 3 张图片被击中
        // picWidth 与 picHeight 分别除以 2,定位击中点为图片的中心
        if (x >= picX1 && x <= picX1 + picWidth && y >= picY1
                && y <= picY1 + picHeight) {
        temp = queue[0];
        queue[0] = queue[2];
        queue[2] = temp;
        findImageView(queue);
    } else if (x >= picX2 && x <= picX2 + picWidth && y >= picY2
                && y <= picY2 + picHeight) {
        temp = queue[2];
        queue[2] = queue[1];
        queue[1] = temp;
        findImageView(queue);
    } else if (x >= picX4 && x <= picX4 + picWidth && y >= picY4
                && y <= picY4 + picHeight) {
        temp = queue[2];
        queue[2] = queue[3];
        queue[3] = temp;
        findImageView(queue);
    } else if (x >= picX5 && x <= picX5 + picWidth && y >= picY5
                && y <= picY5 + picHeight) {
        temp = queue[2];
        queue[2] = queue[4];
        queue[4] = temp;
        findImageView(queue);
    } else if (x >= picX6 && x <= picX6 + picWidth && y >= picY6
                && y <= picY6 + picHeight) {
        temp = queue[2];
        queue[2] = queue[5];
        queue[5] = temp;
        findImageView(queue);
    } else if (x >= picX7 && x <= picX7 + picWidth && y >= picY7
                && y <= picY7 + picHeight) {
        temp = queue[2];
        queue[2] = queue[6];
        queue[6] = temp;
        findImageView(queue);
```

```
            } else if (x >= picX8 && x <= picX8 + picWidth && y >= picY8
                    && y <= picY8 + picHeight) {
                temp = queue[2];
                queue[2] = queue[7];
                queue[7] = temp;
                findImageView(queue);
            } else if (x >= picX9 && x <= picX9 + picWidth && y >= picY9
                    && y <= picY9 + picHeight) {
                temp = queue[2];
                queue[2] = queue[8];
                queue[8] = temp;
                findImageView(queue);
            } else {
            }
        }

        if (clickPic4) {  // 第 4 张图片被击中
            // picWidth 与 picHeight 分别除以 2,定位击中点为图片的中心
            if (x >= picX1 && x <= picX1 + picWidth && y >= picY1
                    && y <= picY1 + picHeight) {
                temp = queue[0];
                queue[0] = queue[3];
                queue[3] = temp;
                findImageView(queue);
            } else if (x >= picX2 && x <= picX2 + picWidth && y >= picY2
                    && y <= picY2 + picHeight) {
                temp = queue[3];
                queue[3] = queue[1];
                queue[1] = temp;
                findImageView(queue);
            } else if (x >= picX3 && x <= picX3 + picWidth && y >= picY3
                    && y <= picY3 + picHeight) {
                temp = queue[2];
                queue[2] = queue[3];
                queue[3] = temp;
                findImageView(queue);
            } else if (x >= picX5 && x <= picX5 + picWidth && y >= picY5
                    && y <= picY5 + picHeight) {
                temp = queue[4];
                queue[4] = queue[3];
                queue[3] = temp;
                findImageView(queue);
```

```
        } else if (x >= picX6 && x <= picX6 + picWidth && y >= picY6
                    && y <= picY6 + picHeight) {
            temp = queue[5];
            queue[5] = queue[3];
            queue[3] = temp;
            findImageView(queue);
        } else if (x >= picX7 && x <= picX7 + picWidth && y >= picY7
                    && y <= picY7 + picHeight) {
            temp = queue[6];
            queue[6] = queue[3];
            queue[3] = temp;
            findImageView(queue);
        } else if (x >= picX8 && x <= picX8 + picWidth && y >= picY8
                    && y <= picY8 + picHeight) {
            temp = queue[7];
            queue[7] = queue[3];
            queue[3] = temp;
            findImageView(queue);
        } else if (x >= picX9 && x <= picX9 + picWidth && y >= picY9
                    && y <= picY9 + picHeight) {
            temp = queue[8];
            queue[8] = queue[3];
            queue[3] = temp;
            findImageView(queue);
        } else {
        }
    }

    if (clickPic5) {    // 第 5 张图片被击中
        // picWidth 与 picHeight 分别除以 2,定位击中点为图片的中心
        if (x >= picX1 && x <= picX1 + picWidth && y >= picY1
                    && y <= picY1 + picHeight) {
            temp = queue[0];
            queue[0] = queue[4];
            queue[4] = temp;
            findImageView(queue);
        } else if (x >= picX2 && x <= picX2 + picWidth && y >= picY2
                    && y <= picY2 + picHeight) {
            temp = queue[4];
            queue[4] = queue[1];
            queue[1] = temp;
            findImageView(queue);
```

```
        } else if (x >= picX3 && x <= picX3 + picWidth && y >= picY3
                && y <= picY3 + picHeight) {
            temp = queue[2];
            queue[2] = queue[4];
            queue[4] = temp;
            findImageView(queue);
        } else if (x >= picX4 && x <= picX4 + picWidth && y >= picY4
                && y <= picY4 + picHeight) {
            temp = queue[4];
            queue[4] = queue[3];
            queue[3] = temp;
            findImageView(queue);
        } else if (x >= picX6 && x <= picX6 + picWidth && y >= picY6
                && y <= picY6 + picHeight) {
            temp = queue[5];
            queue[5] = queue[4];
            queue[4] = temp;
            findImageView(queue);
        } else if (x >= picX7 && x <= picX7 + picWidth && y >= picY7
                && y <= picY7 + picHeight) {
            temp = queue[6];
            queue[6] = queue[4];
            queue[4] = temp;
            findImageView(queue);
        } else if (x >= picX8 && x <= picX8 + picWidth && y >= picY8
                && y <= picY8 + picHeight) {
            temp = queue[7];
            queue[7] = queue[4];
            queue[4] = temp;
            findImageView(queue);
        } else if (x >= picX9 && x <= picX9 + picWidth && y >= picY9
                && y <= picY9 + picHeight) {
            temp = queue[8];
            queue[8] = queue[4];
            queue[4] = temp;
            findImageView(queue);
        } else {
        }
    }

    if (clickPic6) { // 第 6 张图片被击中
        // picWidth 与 picHeight 分别除以 2,定位击中点为图片的中心
```

```
if (x >= picX1 && x <= picX1 + picWidth && y >= picY1
        && y <= picY1 + picHeight) {
    temp = queue[0];
    queue[0] = queue[5];
    queue[5] = temp;
    findImageView(queue);
} else if (x >= picX2 && x <= picX2 + picWidth && y >= picY2
        && y <= picY2 + picHeight) {
    temp = queue[5];
    queue[5] = queue[1];
    queue[1] = temp;
    findImageView(queue);
} else if (x >= picX3 && x <= picX3 + picWidth && y >= picY3
        && y <= picY3 + picHeight) {
    temp = queue[2];
    queue[2] = queue[5];
    queue[5] = temp;
    findImageView(queue);
} else if (x >= picX4 && x <= picX4 + picWidth && y >= picY4
        && y <= picY4 + picHeight) {
    temp = queue[5];
    queue[5] = queue[3];
    queue[3] = temp;
    findImageView(queue);
} else if (x >= picX5 && x <= picX5 + picWidth && y >= picY5
        && y <= picY5 + picHeight) {
    temp = queue[5];
    queue[5] = queue[4];
    queue[4] = temp;
    findImageView(queue);
} else if (x >= picX7 && x <= picX7 + picWidth && y >= picY7
        && y <= picY7 + picHeight) {
    temp = queue[6];
    queue[6] = queue[5];
    queue[5] = temp;
    findImageView(queue);
} else if (x >= picX8 && x <= picX8 + picWidth && y >= picY8
        && y <= picY8 + picHeight) {
    temp = queue[7];
    queue[7] = queue[5];
    queue[5] = temp;
    findImageView(queue);
```

```
    } else if (x >= picX9 && x <= picX9 + picWidth && y >= picY9
                && y <= picY9 + picHeight) {
        temp = queue[8];
        queue[8] = queue[5];
        queue[5] = temp;
        findImageView(queue);
    } else {
    }
}

if (clickPic7) {  // 第 7 张图片被击中
    // picWidth 与 picHeight 分别除以 2,定位击中点为图片的中心

    if (x >= picX1 && x <= picX1 + picWidth && y >= picY1
                && y <= picY1 + picHeight) {
        temp = queue[0];
        queue[0] = queue[6];
        queue[6] = temp;
        findImageView(queue);
    } else if (x >= picX2 && x <= picX2 + picWidth && y >= picY2
                && y <= picY2 + picHeight) {
        temp = queue[6];
        queue[6] = queue[1];
        queue[1] = temp;
        findImageView(queue);
    } else if (x >= picX3 && x <= picX3 + picWidth && y >= picY3
                && y <= picY3 + picHeight) {
        temp = queue[2];
        queue[2] = queue[6];
        queue[6] = temp;
        findImageView(queue);
    } else if (x >= picX4 && x <= picX4 + picWidth && y >= picY4
                && y <= picY4 + picHeight) {
        temp = queue[6];
        queue[6] = queue[3];
        queue[3] = temp;
        findImageView(queue);
    } else if (x >= picX5 && x <= picX5 + picWidth && y >= picY5
                && y <= picY5 + picHeight) {
        temp = queue[6];
        queue[6] = queue[4];
        queue[4] = temp;
```

```
        findImageView(queue);
    } else if (x >= picX6 && x <= picX6 + picWidth && y >= picY6
                && y <= picY6 + picHeight) {
        temp = queue[6];
        queue[6] = queue[5];
        queue[5] = temp;
        findImageView(queue);
    } else if (x >= picX8 && x <= picX8 + picWidth && y >= picY8
                && y <= picY8 + picHeight) {
        temp = queue[7];
        queue[7] = queue[6];
        queue[6] = temp;
        findImageView(queue);
    } else if (x >= picX9 && x <= picX9 + picWidth && y >= picY9
                && y <= picY9 + picHeight) {
        temp = queue[8];
        queue[8] = queue[6];
        queue[6] = temp;
        findImageView(queue);
    } else {
    }
}

if (clickPic8) { // 第 8 张图片被击中
    // picWidth 与 picHeight 分别除以 2,定位击中点为图片的中心
    if (x >= picX1 && x <= picX1 + picWidth && y >= picY1
                && y <= picY1 + picHeight) {
        temp = queue[0];
        queue[0] = queue[7];
        queue[7] = temp;
        findImageView(queue);
    } else if (x >= picX2 && x <= picX2 + picWidth && y >= picY2
                && y <= picY2 + picHeight) {
        temp = queue[7];
        queue[7] = queue[1];
        queue[1] = temp;
        findImageView(queue);
    } else if (x >= picX3 && x <= picX3 + picWidth && y >= picY3
                && y <= picY3 + picHeight) {
        temp = queue[2];
        queue[2] = queue[7];
        queue[7] = temp;
```

```
                findImageView(queue);
            } else if (x >= picX4 && x <= picX4 + picWidth && y >= picY4
                        && y <= picY4 + picHeight) {
                temp = queue[7];
                queue[7] = queue[3];
                queue[3] = temp;
                findImageView(queue);
            } else if (x >= picX5 && x <= picX5 + picWidth && y >= picY5
                        && y <= picY5 + picHeight) {
                temp = queue[7];
                queue[7] = queue[4];
                queue[4] = temp;
                findImageView(queue);
            } else if (x >= picX6 && x <= picX6 + picWidth && y >= picY6
                        && y <= picY6 + picHeight) {
                temp = queue[7];
                queue[7] = queue[5];
                queue[5] = temp;
                findImageView(queue);
            } else if (x >= picX7 && x <= picX7 + picWidth && y >= picY7
                        && y <= picY7 + picHeight) {
                temp = queue[7];
                queue[7] = queue[6];
                queue[6] = temp;
                findImageView(queue);
            } else if (x >= picX9 && x <= picX9 + picWidth && y >= picY9
                        && y <= picY9 + picHeight) {
                temp = queue[8];
                queue[8] = queue[7];
                queue[7] = temp;
                findImageView(queue);
            } else {
            }
        }

        if (clickPic9) { // 第9张图片被击中
            // picWidth与picHeight分别除以2,定位击中点为图片的中心
            if (x >= picX1 && x <= picX1 + picWidth && y >= picY1
                        && y <= picY1 + picHeight) {
                temp = queue[0];
                queue[0] = queue[8];
                queue[8] = temp;
```

```
    findImageView(queue);
} else if (x >= picX2 && x <= picX2 + picWidth && y >= picY2
          && y <= picY2 + picHeight) {
    temp = queue[8];
    queue[8] = queue[1];
    queue[1] = temp;
    findImageView(queue);
} else if (x >= picX3 && x <= picX3 + picWidth && y >= picY3
          && y <= picY3 + picHeight) {
    temp = queue[2];
    queue[2] = queue[8];
    queue[8] = temp;
    findImageView(queue);
} else if (x >= picX4 && x <= picX4 + picWidth && y >= picY4
          && y <= picY4 + picHeight) {
    temp = queue[8];
    queue[8] = queue[3];
    queue[3] = temp;
    findImageView(queue);
} else if (x >= picX5 && x <= picX5 + picWidth && y >= picY5
          && y <= picY5 + picHeight) {
    temp = queue[8];
    queue[8] = queue[4];
    queue[4] = temp;
    findImageView(queue);
} else if (x >= picX6 && x <= picX6 + picWidth && y >= picY6
          && y <= picY6 + picHeight) {
    temp = queue[8];
    queue[8] = queue[5];
    queue[5] = temp;
    findImageView(queue);
} else if (x >= picX7 && x <= picX7 + picWidth && y >= picY7
          && y <= picY7 + picHeight) {
    temp = queue[8];
    queue[8] = queue[6];
    queue[6] = temp;
    findImageView(queue);
} else if (x >= picX8 && x <= picX8 + picWidth && y >= picY8
          && y <= picY8 + picHeight) {
    temp = queue[8];
    queue[8] = queue[7];
    queue[7] = temp;
```

```
                findImageView(queue);
            } else {

            }
        }
        judge();
        break;

    // 移动：当 Action_Down 击中图片时，才执行 move 的动作，否则跳过
    case MotionEvent.ACTION_MOVE:
        if (clickPic1) { // 单击第一张图片
            // picWidth 与 picHeight 分别除以 2,定位击中点为图片的中心
            paramsIv_1.x = (int) (x - picWidth / 2);
            paramsIv_1.y = (int) (y - picHeight / 2);
            // 通过 setLayoutParams 达到图片跟随手指移动的目的
            iv_1.setLayoutParams(paramsIv_1);
        }

        if (clickPic2) {
            // picWidth 与 picHeight 分别除以 2,定位击中点为图片的中心
            paramsIv_2.x = (int) (x - picWidth / 2);
            paramsIv_2.y = (int) (y - picHeight / 2);
            iv_2.setLayoutParams(paramsIv_2);
        }

        if (clickPic3) {
            // picWidth 与 picHeight 分别除以 2,定位击中点为图片的中心
            paramsIv_3.x = (int) (x - picWidth / 2);
            paramsIv_3.y = (int) (y - picHeight / 2);
            iv_3.setLayoutParams(paramsIv_3);
        }

        if (clickPic4) {
            // picWidth 与 picHeight 分别除以 2,定位击中点为图片的中心
            paramsIv_4.x = (int) (x - picWidth / 2);
            paramsIv_4.y = (int) (y - picHeight / 2);
            iv_4.setLayoutParams(paramsIv_4);
        }

        if (clickPic5) {
            // picWidth 与 picHeight 分别除以 2,定位击中点为图片的中心
            paramsIv_5.x = (int) (x - picWidth / 2);
```

```
            paramsIv_5.y = (int) (y - picHeight / 2);
            iv_5.setLayoutParams(paramsIv_5);
        }

        if (clickPic6) {
            // picWidth 与 picHeight 分别除以 2,定位击中点为图片的中心
            paramsIv_6.x = (int) (x - picWidth / 2);
            paramsIv_6.y = (int) (y - picHeight / 2);
            iv_6.setLayoutParams(paramsIv_6);
        }

        if (clickPic7) {
            // picWidth 与 picHeight 分别除以 2,定位击中点为图片的中心
            paramsIv_7.x = (int) (x - picWidth / 2);
            paramsIv_7.y = (int) (y - picHeight / 2);
            iv_7.setLayoutParams(paramsIv_7);
        }

        if (clickPic8) {
            // picWidth 与 picHeight 分别除以 2,定位击中点为图片的中心
            paramsIv_8.x = (int) (x - picWidth / 2);
            paramsIv_8.y = (int) (y - picHeight / 2);
            iv_8.setLayoutParams(paramsIv_8);
        }

        if (clickPic9) {
            // picWidth 与 picHeight 分别除以 2,定位击中点为图片的中心
            paramsIv_9.x = (int) (x - picWidth / 2);
            paramsIv_9.y = (int) (y - picHeight / 2);
            iv_9.setLayoutParams(paramsIv_9);
        }
        break;
    }
    return super.onTouchEvent(event);
}
```

最后,我们要时刻检查图片是否拼成功,刚开始设计项目的时候我们分别把打乱了的图片和没打乱的图片存到了两个数组里,现在通过两个数组的比较就能检查图片是否拼成功了,代码如下:

```
/**
  * 图片匹配成功
  **/
```

```
public void judge() {
    if (queue[0].equals(queue1[0]) && queue[1].equals(queue1[1])
            && queue[2].equals(queue1[2]) && queue[3].equals(queue1[3])
            && queue[4].equals(queue1[4]) && queue[5].equals(queue1[5])
            && queue[6].equals(queue1[6]) && queue[7].equals(queue1[7])
            && queue[8].equals(queue1[8])) {
        Toast.makeText(this, "恭喜你,拼图完成!!", Toast.LENGTH_SHORT).show();
        flag = true;

    }
}
```

第七步,选取本地拼图并进行拼图游戏已经完成,接下来我们看看自己制作新的拼图是怎么实现的,其实代码很简单,只需调用一下系统的照相机,然后得到所拍的照片就可以了。

具体实现过程如下:

```
public class MyCamera extends Activity {
  private static final int CAPTURE_IMAGE_ACTIVITY_REQUEST_CODE = 100;
  private Uri fileUri;
  public static final int MEDIA_TYPE_IMAGE = 1;
  public static final int MEDIA_TYPE_VIDEO = 2;

  @Override
  protected void onCreate(Bundle savedInstanceState) {
      super.onCreate(savedInstanceState);
      setContentView(R.layout.camera);
        //调用系统的照相机
      Intent intent = new Intent(MediaStore.ACTION_IMAGE_CAPTURE);
      intent.putExtra(MediaStore.EXTRA_OUTPUT, fileUri);
      startActivityForResult(intent, CAPTURE_IMAGE_ACTIVITY_REQUEST_CODE);
  }

      @Override
      protected void onActivityResult(int requestCode, int resultCode, Intent data) {
          super.onActivityResult(requestCode, resultCode, data);
            // 检查手机是否安装了 SD 卡
          if(requestCode == 100){
                if (resultCode == Activity.RESULT_OK) {
                      String sdStatus = Environment.getExternalStorageState();
                      if (! sdStatus.equals(Environment.MEDIA_MOUNTED)) {
                            Log.v("Wedding",
                                        "SD card id not avaiable/writeable right now.");
```

```
                return;
            }
              //得到所照的照片
            Bundle bundle = data.getExtras();
            Bitmap bitmap = (Bitmap) bundle.get("data");
              //将所照的照片切割成横向3份,纵向3份
            ImageSplitter.split(bitmap, 3, 3);
              //跳转到Puzzle类,完成拼图游戏
            Intent intent = new Intent(this,Puzzle.class);
            startActivity(intent);

        }
        }
    }

}
```

8.6　本章小结

本章通过项目案例介绍了一款拼图游戏的设计与实现过程,其核心功能是图片切割、打乱顺序与图片排序。其中,图片切割利用了 Android 平台封装好的 ImageSplitter 类和 ImagePiece 类;而打乱图片顺序,则是利用了随机数原理;图片排序,是通过对手指触摸屏的处理,分别记录图片的横、纵坐标,经过一系列的算法与位置的移动实现图片的移动与拼接。

第9章 蝴蝶飞飞游戏项目案例

蝴蝶飞飞游戏是一款休闲娱乐软件，它属于游戏类软件。本章以蝴蝶飞飞游戏项目为例介绍游戏类软件项目 UI 的一般设计方法，以及本软件所涉及的游戏算法、游戏关卡、游戏信息、背景音乐等相关内容。

9.1 预备知识

(1) Service 服务

Android 四大组件之一，无界面，后台运行，不是一个线程，是程序的一部分，不能与用户交互，不能自己启动，使用 startService 和 bindService 来启动服务。本章使用 service 来后台播放背景音乐。

(2) MotionEvent 触摸事件

当用户触摸屏幕时将创建一个 MotionEvent 对象。MotionEvent 包含关于发生触摸的位置和时间等细节信息。

(3) OnTouch 函数

```
public boolean onTouch(View v, MotionEvent event)
```

该函数的返回值是 boolean，原因是系统对这些事件的处理是有条件的，必须满足条件才会触发相应的事件处理函数。onTouch 如果返回 true，则表明对该事件做了处理。

(4) Handler 线程机制

主要接收子线程发送数据来配合主线程更新 UI，可以分发 menssage 或者 runnable 对象到主线程中，每个 Handler 都会绑定到创建它的线程中(一般是主线程)。本章使用 Handler 来更新游戏界面 UI。

(5) Canvas 画布

把要画的东西画到画布上，此时就用 Canvas 类。本章中使用 Canvas 来做游戏场景的绘制。

(6) SurfaceView

SurfaceView 是视图(View)的继承类，这个视图里内嵌了一个专门用于绘制的 Surface。可以控制这个 Surface 的格式和尺寸。SurfaceView 控制这个 Surface 的绘制位置。

9.2　游戏需求分析

本游戏软件以休闲、“萌”为主题。在游戏中，一只华美的蝴蝶翩翩而至，好似那跳动的精灵，起舞在花丛中，寻找理想的花朵儿栖息，然而它的面前有不少的毒虫会“抓”住它。本游戏的玩法就是：让用户身临其境，控制蝴蝶行进的方向，躲过毒虫的侵扰；用户只需要一只手指在蝴蝶飞行路径的前方画上一道线，蝴蝶碰到这道障碍物就会改变行迹路线；用户经过一系列操作，控制蝴蝶最终到达理想的花朵，完成目标通关。游戏中设置了许多关卡，一级级增设了游戏的难度，每个阶段都会有不同的变化；同时，游戏中简洁的主题，清新的画面，怒放的花朵，灵动的蝴蝶配以素雅的音乐，这一切都会让用户更加享受这一妙趣横生的小游戏带来的乐趣，让闲暇时身心得以放松。

9.3　功能分析

作为一款闯关类游戏，这款手机游戏软件操作方法与以往不同，它既不是采用重力感应也不是采用方向按键直接控制目标，而是采用画线来反弹目标，这需要计算一定的角度，因此，它比一般游戏更具有难度和挑战。此外，这款游戏还具有高分榜和分享功能，能够将游戏得分发布到所有人，与朋友们一起分享。

9.4　设　　计

蝴蝶飞飞游戏软件的界面设计与功能描述如表 9-1 所示。

表 9-1　蝴蝶飞飞游戏软件的界面设计与功能描述

界面	功能简述
主界面	1. 显示信息 游戏名称 2. 功能按钮 (1) 开始 (2) 帮助 (3) 设置 (4) 退出 (5) 关于 (6) 分享
选择关卡界面	选择不同的关卡，关卡不同，难度不同
游戏界面	通过手指的触控使蝴蝶避过害虫到达花朵处

9.4.1 UI设计

(1) 主界面如图 9-1 所示，功能包括开始、帮助、设置、退出、关于、分享功能。

图 9-1　主界面

(2) 选择关卡界面如图 9-2 所示。

图 9-2　关卡选择

(3) 游戏界面如图 9-3 所示，功能包括显示时间、显示分数、暂停游戏。

(4) 游戏设置界面如图 9-4 所示。

(5) 关于本项目的介绍界面如图 9-5 所示。

图 9-3　游戏界面

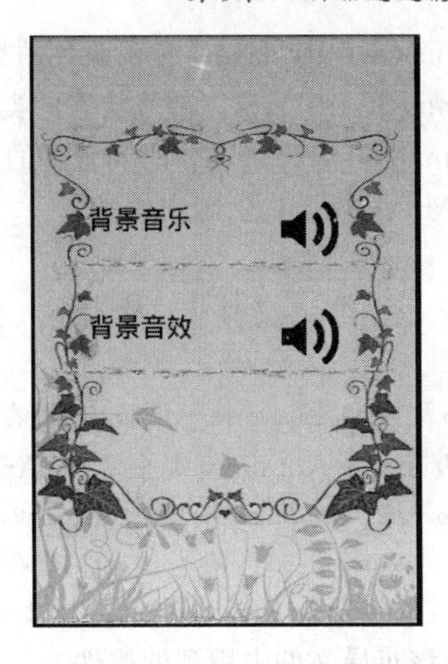

图 9-4　游戏设置界面

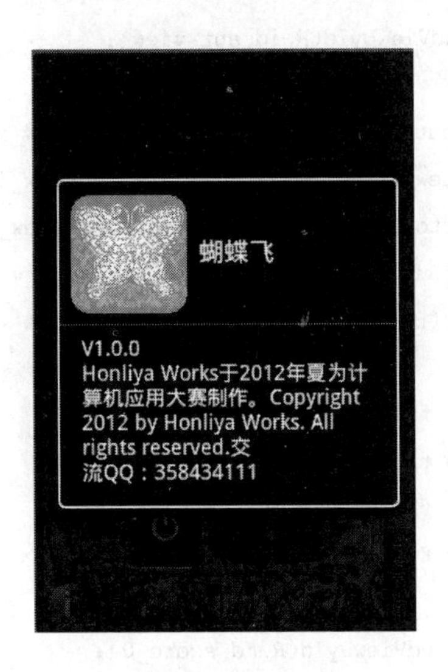

图 9-5　介绍界面

9.4.2　类设计

软件中用到的类如下：

(1) Splash.java。程序启动时 Logo 界面的类。

(2) MainPage.java。主界面(选择开始、帮助、设置、退出)的类。

(3) LevelActivity.java。选择关卡的类。

(4) AntGuideActivity.java。游戏开始界面的类。

(5) GamePref.java。用于保存用户游戏信息的类。

(6) MusicReceiver.java。音乐广播的类。

(7) MusicService. java。音乐服务的类。

(8) GameStatus. java。游戏状态的类。

(9) TimeManager. java。时间管理的类。

(10) AntView. java。画布的类。

9.5 编码实现

游戏布局文件是 game_view. xml,在该文件中有这样一个视图,它是一个自定义 View。此 View 由 AntView 类来实现,AntView 类用于创建一个游戏视图。

```
<com.howfun.android.antguide.view.AntView
        android:id="@+id/ant_view" android:layout_width="fill_parent"
        android:layout_height="fill_parent" />
```

获取在该布局文件中用到的控件。

```
gamePause = (ImageView) findViewById(R.id.game_pause);
gamePlay = (ImageView) findViewById(R.id.game_play);
antView = (AntView) findViewById(R.id.ant_view);

mGameInfo = (LinearLayout) findViewById(R.id.game_view_info);
mGameInfoText = (TextView) findViewById(R.id.game_view_info_text);
mGameInfoPlayBtn = (Button) findViewById(R.id.game_view_play_btn);
mGameInfoRestartBtn = (Button) findViewById(R.id.game_view_restart_btn);
mGameInfoNextLvBtn =   (Button) findViewById(R.id.game_view_next_level);

mTimeMin0 = (ImageView) findViewById(R.id.time_min0);
mTimeMin1 = (ImageView) findViewById(R.id.time_min1);
mTimeSec0 = (ImageView) findViewById(R.id.time_sec0);
mTimeSec1 = (ImageView) findViewById(R.id.time_sec1);

mScore0 = (ImageView) findViewById(R.id.score_0);
mScore1 = (ImageView) findViewById(R.id.score_1);
mScore2 = (ImageView) findViewById(R.id.score_2);
```

设置监听。

根据单击不同游戏跳转可以实现:跳转下一级别、开始游戏、暂停游戏、重新开始游戏、暂停游戏和继续游戏。

```
if (mGameInfoNextLvBtn != null) {
        mGameInfoNextLvBtn.setOnClickListener(new OnClickListener() {
            @Override
            public void onClick(View v) {
                goNextLv();
                resetGame();
```

```
            }
        });
    }

    if (mGameInfoPlayBtn != null) {
        mGameInfoPlayBtn.setOnClickListener(new OnClickListener() {
            @Override
            public void onClick(View v) {
                if (mGameStatus.getStatus() == GameStatus.GAME_PAUSED) {
                    resumeGame();
                }
            }
        });
    }

    if (mGameInfoRestartBtn != null) {
        mGameInfoRestartBtn.setOnClickListener(new OnClickListener() {
            @Override
            public void onClick(View v) {
                resetGame();
            }
        });
    }

    if (gamePause != null) {
        gamePause.setOnClickListener(new OnClickListener() {
            @Override
            public void onClick(View v) {
                pauseGame();
            }
        });
    }

    if (gamePlay != null) {
        gamePlay.setOnClickListener(new OnClickListener() {
            @Override
            public void onClick(View v) {
                if (mGameStatus.getStatus() == GameStatus.GAME_PAUSED) {
                    resumeGame();
                }
            }
        });
```

```
        }
```

加载触摸监听，返回触摸事件，将 X，Y 赋值给 x，y，并重写 Ontouch 函数，根据触摸过程中的手指按下、滑动、抬起 3 个状态画出游戏中的“挡板”即改变虫子轨迹的线。返回 true 时 event 有效。

```
if (antView != null) {
            antView.setOnTouchListener(new OnTouchListener() {

                @Override
                public boolean onTouch(View v, MotionEvent event) {
                    int action = event.getAction();
                    float x = event.getX();
                    float y = event.getY();
                    switch (action) {
                    case MotionEvent.ACTION_DOWN:
                        antView.setDownPos(x, y);
                        break;
                    case MotionEvent.ACTION_MOVE:
                        break;
                    case MotionEvent.ACTION_UP:
                        antView.setUpPos(x, y);
                        antView.showBlockLine();
                        break;
                    }
                    return true;
                }

            });
        }
    }
```

(1) 游戏状态初始化

```
private void init() {
    //初始化记分牌,单态,只有此一个分数实例
    mScore = GamePref.getInstance(this).getScorePref();
    initScoreBoard();
    Utils.log(TAG, "score is " + mScore);
    //初始化音效,背景音
    mIntentService = new Intent("com.howfun.android.antguide.MusicService");
    mIntentReceiver = new Intent("com.howfun.android.antguide.MusicReceiver");
    //初始化游戏状态,并建立新的时间子线程
    mGameStatus = new GameStatus();
```

```
        mTimeManager = new TimeManager(mHandler);
        //加载设置布局
        mSettings = getSharedPreferences(Utils.PREF_SETTINGS, 0);
        mIsBackPressed = false;
}
```

(2) LevelActivity 设计(实现了游戏关卡的选择)

属性设置如下:

```
private static final String TAG = "LevelActivity";
private Gallery mGallery = null;
private LevelGalleryAdapter mAdapter;
private int mPassedLevel;
private Context mContext;
public void onCreate(Bundle savedInstanceState) {
        super.onCreate(savedInstanceState);
        getWindow().setFlags(WindowManager.LayoutParams.FLAG_FULLSCREEN,
                WindowManager.LayoutParams.FLAG_FULLSCREEN);
        requestWindowFeature(Window.FEATURE_NO_TITLE);
        setContentView(R.layout.level_select);

        mContext = this;

        mPassedLevel = GamePref.getInstance(this).getLevelPref();
        Utils.log(TAG, "passed game level = " + mPassedLevel);

        mGallery = (Gallery) findViewById(R.id.level_select_gallery);
        mAdapter = new LevelGalleryAdapter(this);
        mGallery.setAdapter(mAdapter);
        mGallery.setSpacing(30);

        mGallery.setOnItemClickListener(new OnItemClickListener() {

            @Override
            public void onItemClick(AdapterView<?> parent, View view,
                    int position, long id) {

                if (getClickable(position)) {
                    Intent intent = new Intent(LevelActivity.this,
                            AntGuideActivity.class);
                    intent.putExtra(Utils.LEVEL_REF, position);
                    startActivity(intent);
                }else {
```

```
                Toast.makeText(mContext, R.string.lock, Toast.LENGTH_LONG).show();
            }

        }
    });
    Utils.setAD(this);
}

public void onResume() {
    super.onResume();

    if (mAdapter != null) {
        Utils.log(TAG, "back to levell activity");
        mPassedLevel = GamePref.getInstance(this).getLevelPref();
        mAdapter.notifyDataSetChanged();
    }

}

public class LevelGalleryAdapter extends BaseAdapter {
    private Context mContext;

    public LevelGalleryAdapter(Context context) {
        mContext = context;
    }

    public int getCount() {
        return AntMap.MAP_COUNT;
    }

    public Object getItem(int position) {
        return position;
    }

    public long getItemId(int position) {
        return position;
    }

    public View getView(int position, View convertView, ViewGroup parent) {
        LayoutInflater inflater = LayoutInflater.from(mContext);
        View ant = inflater.inflate(R.layout.level_gallery_item, null);
```

```
        CustomNumberView text = (CustomNumberView) ant.findViewById
                                                (R.id.level_gallery_item_level);
        text.setNum(position + 1);

        ImageView passView = (ImageView) ant
                .findViewById(R.id.level_gallery_item_watermark);
        //if passed:
        boolean isPassed = mPassedLevel >= position;
        if (isPassed) {
            //set unlock icon
            passView.setImageResource(R.drawable.passed);
        } else {
            //if is next level of passed one, set unlock icon
            //else set lock icon,

            boolean nextLevel = (mPassedLevel + 1 == position );
            if (nextLevel) {
                passView.setImageResource(R.drawable.new_icon);
            } else {
                passView.setBackgroundResource(R.drawable.lock);
            }
        }

        return ant;
    }
}
private boolean getClickable(int position) {
    if (position <= mPassedLevel + 1) {
        return true;
    } else {
        return false;
    }
}
```

(3) AntGuideActivity 设计(开始游戏界面设置)

所需属性设置如下:

```
private static final String TAG = "AntGuide";

    private static final String PREF = "ant pref";
    private static final String GAME_STATE_PREF = "ant state pref";// Paused is 1
        //控制选择声音
    private Sound mSound;
```

```
//场景大小
public static int DEVICE_WIDTH;
public static int DEVICE_HEIGHT;
//定义布尔型变量,判断是否按下返回
private boolean mIsBackPressed;
//5种声音变量
private static final int SOUND_EFFECT_COLLISION = 0;
private static final int SOUND_EFFECT_FOOD = 1;
private static final int SOUND_EFFECT_VICTORY = 2;
private static final int SOUND_EFFECT_LOST = 3;
private static final int SOUND_EFFECT_TRAPPED = 4;
//最大分数
private static final int MAX_SCORE = 999;
// 9种数字图片
private static final int[] nums = { R.drawable.num_0, R.drawable.num_1,
        R.drawable.num_2, R.drawable.num_3, R.drawable.num_4,
        R.drawable.num_5, R.drawable.num_6, R.drawable.num_7,
        R.drawable.num_8, R.drawable.num_9 };

private SoundPool mSoundPool;//用以选定声音
private int[] mSounds;  //声音数组
private int[] mSoundIds; //用以接收声音标签

Intent mIntentService = null;
Intent mIntentReceiver = null;

// 图片按钮开始,暂停
private ImageView gamePause;
private ImageView gamePlay;

private AntView antView;
private LinearLayout mGameInfo;  //游戏布局
private TextView mGameInfoText;  // 休息提示字
//时间图片
private ImageView mTimeMin0;
private ImageView mTimeMin1;
private ImageView mTimeSec0;
private ImageView mTimeSec1;

private ImageView mScore0;
private ImageView mScore1;
private ImageView mScore2;
```

```
private int mScore = 0;

private GameStatus mGameStatus;
private TimeManager mTimeManager;

private SharedPreferences mSettings;
private boolean mBackMusicOff;  //布尔型,设置背景音乐开关
private boolean mSoundEffectOff; //布尔型 ,设置音效开关
```

(4) 游戏开始的设置方法

```
/*
* 继续游戏监听,根据单击游戏的不同跳转至下一级别,开始游戏。
* 暂停游戏,重新开始游戏;暂停游戏,继续游戏。
*/
  private void setupListeners() {
      if (mGameInfoNextLvBtn != null) {
        mGameInfoNextLvBtn.setOnClickListener(new OnClickListener() {
            @Override
            public void onClick(View v) {
              goNextLv();
              resetGame();
            }
        });
      }

      if (mGameInfoPlayBtn != null) {
        mGameInfoPlayBtn.setOnClickListener(new OnClickListener() {
            @Override
            public void onClick(View v) {
              if (mGameStatus.getStatus() == GameStatus.GAME_PAUSED) {
                  resumeGame();
              }
            }
        });
      }
      if (mGameInfoRestartBtn != null) {
        mGameInfoRestartBtn.setOnClickListener(new OnClickListener() {
            @Override
            public void onClick(View v) {
              resetGame();
            }
        });
```

```
        }

        if (gamePause != null) {
            gamePause.setOnClickListener(new OnClickListener() {
                @Override
                public void onClick(View v) {
                    pauseGame();
                }
            });
        }

        if (gamePlay != null) {
            gamePlay.setOnClickListener(new OnClickListener() {
                @Override
                public void onClick(View v) {
                    if (mGameStatus.getStatus() == GameStatus.GAME_PAUSED) {
                        resumeGame();
                    }
                }
            });
        }
    /*
     * 加载触摸监听,返回触摸事件,将X,Y赋值给x,y,并重写Ontouch函数。
     * 根据触摸过程中的手指按下、滑动、抬起3个状态画出游戏中的“挡板”
     * 即改变虫子轨迹的线。返回true时event有效
     */
        if (antView != null) {
            antView.setOnTouchListener(new OnTouchListener() {

                @Override
                public boolean onTouch(View v, MotionEvent event) {
                    int action = event.getAction();
                    float x = event.getX();
                    float y = event.getY();
                    switch (action) {
                    case MotionEvent.ACTION_DOWN:
                        antView.setDownPos(x, y);

                        break;
                    case MotionEvent.ACTION_MOVE:
                        break;
                    case MotionEvent.ACTION_UP:
```

```
                antView.setUpPos(x, y);
                antView.showBlockLine();
                break;
            }
            return true;
        }

    });
}

}
//采用单态,以保证只有此一个实例。所以用 getinstance 实现
protected void goNextLv() {

    if (mScore>0) {
        GamePref.getInstance(this).SetScorePref(mScore);
        Utils.log(TAG, "write score to pref: " + mScore);
    }

    if (antView != null) {
        antView.goNextLv();
    }
}
//从音效数组中加载音效到"音池"
private void loadSoundEffects() {
    mSounds = new int[] { R.raw.collision, R.raw.food, R.raw.victory,
            R.raw.lost, R.raw.trapped };
    mSoundPool = new SoundPool(mSounds.length, AudioManager.STREAM_MUSIC, 100);
    mSoundIds = new int[] {
            mSoundPool.load(this, mSounds[SOUND_EFFECT_COLLISION], 1),
            mSoundPool.load(this, mSounds[SOUND_EFFECT_FOOD], 1),
            mSoundPool.load(this, mSounds[SOUND_EFFECT_VICTORY], 1),
            mSoundPool.load(this, mSounds[SOUND_EFFECT_LOST], 1),
            mSoundPool.load(this, mSounds[SOUND_EFFECT_TRAPPED], 1) };
}

private void playSoundEffect(int id) {
    if (mSoundEffectOff)
        return;
    if (mSoundPool != null) {
        mSoundPool.play(mSoundIds[id], 13, 15, 1, 0, 1f);
    }
```

```
    }
  //游戏状态初始化
    private void init() {
        //初始化记分牌,单态,只有此一个分数实例
        mScore = GamePref.getInstance(this).getScorePref();
        initScoreBoard();

        Utils.log(TAG, "score is " + mScore);
        //初始化音效,背景音
        mIntentService = new Intent("com.howfun.android.antguide.MusicService");
        mIntentReceiver = new Intent("com.howfun.android.antguide.MusicReceiver");
        //初始化游戏状态,并建立新的时间子线程
        mGameStatus = new GameStatus();
        mTimeManager = new TimeManager(mHandler);
        //加载设置布局
        mSettings = getSharedPreferences(Utils.PREF_SETTINGS, 0);
        mIsBackPressed = false;
    }
     //更新游戏状态至运行
    private void playGame() {
        Utils.log(TAG, "playGame..");
        mGameStatus.setStaus(GameStatus.GAME_RUNNING);
      //在场景中显示暂停按钮
        showGamePause();
        hideGameInfo();
       //时间清零并开始计时
        mTimeManager.reset();
        mTimeManager.start();
    }
     //重置游戏,重新加载虫子,跳转"开始游戏"
    private void resetGame() {
        Utils.log(TAG, "reset game");
        antView.resetGame();
        playGame();
    }
    //继续游戏,更新游戏状态,恢复虫子,继续计时
    private void resumeGame() {
        mGameStatus.setStaus(GameStatus.GAME_RUNNING);
        antView.resumeGame();

        showGamePause();
        hideGameInfo();
```

```
        mTimeManager.resume();
    }

    /*
 * 按下暂停键后,暂停游戏 activity,更新游戏状态至暂停,隐藏暂停键,
 * 显示继续键,暂停计时器
     */
    private void pauseGame() {
        Utils.log(TAG, "pauseGame..");
        antView.pauseGame();
        mGameStatus.setStaus(GameStatus.GAME_PAUSED);
        hideGamePause();
        showGameInfo(Utils.ANT_PAUSED, R.string.paused);
        // TODO timing pause
        mTimeManager.pause();
    }

    /*
     * 当虫子到家,走出场景丢失,被困住,超时时终止游戏
     */
    private void stopGame(int why) {
        mGameStatus.setStaus(GameStatus.GAME_STOPPED);
        antView.stopGame();
        hideGamePause();

        int info = R.string.app_name;

        if (why == Utils.ANT_HOME) {
            info = R.string.get_home;
        } else if (why == Utils.ANT_LOST) {
            info = R.string.lost;
        } else if (why == Utils.ANT_TIMEOUT) {
            info = R.string.time_out;
        } else if (why == Utils.ANT_TRAPPED) {
            info = R.string.trapped;
        }
        else {
            info = R.string.app_name;
        }
        showGameInfo(why, info);
        mTimeManager.stop();
```

```
        // if (isHighScore(mScore)) {
        // Message msg = new Message();
        // msg.what = Utils.MSG_SCORE_BOARD;
        // msg.arg1 = mScore;
        // mHandler.sendMessage(msg);
        // }
    }
//显示暂停键
private void showGamePause() {
        gamePause.setVisibility(View.VISIBLE);
        gamePlay.setVisibility(View.INVISIBLE);
    }
//隐藏暂停键
private void hideGamePause() {
        gamePause.setVisibility(View.INVISIBLE);
//      gamePlay.setVisibility(View.VISIBLE);
    }

private void hideGameInfo() {
        mGameInfo.setVisibility(View.GONE);
    }

/*
 * 游戏信息的更新依据以及场景跳转
 */
private void showGameInfo(int why, int infoId) {

        String info = this.getResources().getString(infoId);
        mGameInfoText.setText(info);
        mGameInfo.setVisibility(View.VISIBLE);
        switch (why) {
        case Utils.ANT_HOME:              //虫子到家后,显示下一级别提示
            mGameInfoPlayBtn.setVisibility(View.GONE);
            mGameInfoRestartBtn.setVisibility(View.GONE);
            mGameInfoNextLvBtn.setVisibility(View.VISIBLE);
            break;
        case Utils.ANT_LOST:              //虫子走丢,显示重新开始
            mGameInfoPlayBtn.setVisibility(View.GONE);
            mGameInfoRestartBtn.setVisibility(View.VISIBLE);
            mGameInfoNextLvBtn.setVisibility(View.GONE);
            break;
        case Utils.ANT_PAUSED:            //暂停时,显示"继续"按键
```

```
            mGameInfoPlayBtn.setVisibility(View.VISIBLE);
            mGameInfoRestartBtn.setVisibility(View.GONE);
            mGameInfoNextLvBtn.setVisibility(View.GONE);
            break;
        case Utils.ANT_TRAPPED:        //虫子被困住,显示"重新开始"按键
            mGameInfoPlayBtn.setVisibility(View.GONE);
            mGameInfoRestartBtn.setVisibility(View.VISIBLE);
            mGameInfoNextLvBtn.setVisibility(View.GONE);
            break;
        }
    }
    @Override
    public boolean onTouch(View v, MotionEvent event) {
        return false;
    }
/*
 * 利用取余法获取分数,a 对 10 取余,余数即为个位数,除以 10 再对 10 取余,
 * 余数即为十位, 除 100,再对 10 取余即为百位,分别存于分数数组之中
 */
    private int getUnit(int score) {
        int unit;
        if (score<10) {
            unit = score;
        } else if (score<100) {
            unit = score % 10;
        } else {
            int temp = score % 100;
            unit = temp % 10;
        }
        return unit;
    }

    private int getTen(int score) {
        int ten;
        if (score<10) {
            ten = 0;
        } else if (score<100) {
            ten = score / 10;
        } else {
            int temp = score % 100;
            ten = temp / 10;
        }
```

```
        return ten;
    }

    private int getHundred(int score) {
        int hundred;
        if (score<100) {
            hundred = 0;
        } else {
            hundred = score / 100;
        }
        return hundred;
    }
    //设置记分牌
    private void initScoreBoard() {

        int score0 = getHundred(mScore);
        int score1 = getTen(mScore);
        int score2 = getUnit(mScore);
        mScore0.setBackgroundResource(nums[score0]);
        mScore1.setBackgroundResource(nums[score1]);
        mScore2.setBackgroundResource(nums[score2]);
    }
  //更新记分牌
    private void updateScore() {
        int score = mScore;
        mScore++;
       if (mScore>MAX_SCORE)
            return;
        int score0 = getHundred(mScore);
        int score1 = getTen(mScore);
        int score2 = getUnit(mScore);

        int score0t = getHundred(score);
        int score1t = getTen(score);
        boolean tenChange = false;
        boolean hundredChange = false;
        if (score0 != score0t)
            hundredChange = true;
        if (score1 != score1t)
            tenChange = true;
      //利用 setAnimation 加载分数变化动画,anim 为加载条件
        mScore2.setAnimation(AnimationUtils
```

```
                .loadAnimation(this, R.anim.push_up_in));
        mScore2.setBackgroundResource(nums[score2]);
        mScore2.setAnimation(AnimationUtils.loadAnimation(this,
                R.anim.push_up_out));
        if (tenChange) {
            mScore1.setAnimation(AnimationUtils.loadAnimation(this,
                    R.anim.push_up_in));
            mScore1.setBackgroundResource(nums[score1]);
            mScore1.setAnimation(AnimationUtils.loadAnimation(this,
                    R.anim.push_up_out));
        }
        if (hundredChange) {
            mScore0.setAnimation(AnimationUtils.loadAnimation(this,
                    R.anim.push_up_in));
            mScore0.setBackgroundResource(nums[score0]);
            mScore0.setAnimation(AnimationUtils.loadAnimation(this,
                    R.anim.push_up_out));
        }

    }

    private void showScoreBoard(long score) {
        Intent intent = new Intent(this, BigNameActivity.class);
        intent.putExtra(Utils.SCORE, score);
        startActivity(intent);
    }

    private void updateTimeBoard(String time) {
        int min0 = Integer.parseInt(String.valueOf(time.charAt(0)));
        int min1 = Integer.parseInt(String.valueOf(time.charAt(1)));
        int sec0 = Integer.parseInt(String.valueOf(time.charAt(2)));
        int sec1 = Integer.parseInt(String.valueOf(time.charAt(3)));
        mTimeMin0.setBackgroundResource(nums[min0]);
        mTimeMin1.setBackgroundResource(nums[min1]);
        mTimeSec0.setBackgroundResource(nums[sec0]);
        mTimeSec1.setBackgroundResource(nums[sec1]);
    }

    private boolean isHighScore(long score) {
        boolean flag = false;
        if (score == 0) {
            return false;
```

```
    }
/*
*从数据库取出最高分数,与当前分数比较,若是新最高分数,
*刷新数据库,否则抛弃,并返回标记
*/
        List<Score> l = new ArrayList<Score>();
        DBAdapter db = new DBAdapter(this);
        db.open();
        l = db.getHighScores(Utils.TOP_SCORE_COUNT);
        db.close();
        if (l.size()<Utils.TOP_SCORE_COUNT) {
            flag = true;
        } else {
            long scoreT = l.get(l.size() - 1).getScore();
            if (score >= scoreT) {
                flag = true;
            }
        }
        return flag;
    }
    @Override
    public boolean onKeyDown(int keyCode, KeyEvent event) {
        Utils.log(TAG, "onKeyDown");
        switch (keyCode) {
        case KeyEvent.KEYCODE_BACK:
            mIsBackPressed = true;
            antView.pauseGame();
            mGameStatus.setStaus(GameStatus.GAME_PAUSED);
            break;
        default:
            break;
        }
        return super.onKeyDown(keyCode, event);
    }

    @Override
    protected void onSaveInstanceState(Bundle outState) {
        Utils.log(TAG, "onSaveInstanceState");
    }

    //存储游戏状态
    protected void onRestoreInstanceState(Bundle outState) {
```

```
        Utils.log(TAG, "onRestoreInstanceState");
        SharedPreferences sp = this.getSharedPreferences(PREF, MODE_PRIVATE);
        int status = sp.getInt(GAME_STATE_PREF, GameStatus.GAME_INIT);
        if (status == GameStatus.GAME_PAUSED) {
            restorePausedGame();
        }
}
    //暂停状态
    private void resetState() {
        SharedPreferences sp = this.getSharedPreferences(PREF, MODE_PRIVATE);
        sp.edit().putInt(GAME_STATE_PREF, GameStatus.GAME_INIT).commit();

    }
    private void saveState() {
        Utils.log(TAG, "saveState");
        antView.getGameStatus(mGameStatus);
        SharedPreferences sp = this.getSharedPreferences(PREF, MODE_PRIVATE);
        sp.edit().putInt(GAME_STATE_PREF, GameStatus.GAME_PAUSED).commit();
        float x = mGameStatus.getAntPos().x;
        float y = mGameStatus.getAntPos().y;
        float angle = mGameStatus.getAntAngle();
        sp.edit().putFloat("x", x).commit();
        sp.edit().putFloat("y", y).commit();
        sp.edit().putFloat("angle", angle).commit();
        sp.edit().putInt("time", this.mTimeManager.getTime()).commit();
    }
    private void restorePausedGame() {
    SharedPreferences sp = this.getSharedPreferences(PREF, MODE_PRIVATE);
        float x = sp.getFloat("x", 0);
        float y = sp.getFloat("y", 0);
        float angle = sp.getFloat("angle", 0);
        int time = sp.getInt("time", 0);
        if (mTimeManager != null) {
            mTimeManager.restoreTime(time);
            mTimeManager.pause();
        } else {
            Utils.log(TAG, "llllllllllllllltime is null");
        }
      mGameStatus.setStaus(GameStatus.GAME_PAUSED);
        mGameStatus.setAntAngle(angle);
        mGameStatus.setAntPos(new Pos(x, y));
        if (antView != null) {
```

```
        antView.setRestoredState(mGameStatus);
    }
  }
```

(5) ViewPaperActivity 设计(帮助界面设置)

此类中使用了 ViewPager,此时需要导入一个 jar 包,才能使用此类。

```
private ViewPager vp;
private ViewPaperAdapter vpAdapter;
private List<View> views;
//引导图片资源
private static final int[] pics = { R.drawable.helppage_1,
R.drawable.helppage_2, R.drawable.helppage_3 };
//底部小点图片
  private ImageView[] dots ;
  //记录当前选中位置
  private int currentIndex;
views = new ArrayList<View>();
LinearLayout.LayoutParams mParams = new LinearLayout.LayoutParams
        (LinearLayout.LayoutParams.WRAP_CONTENT,
        LinearLayout.LayoutParams.WRAP_CONTENT);
    //初始化引导图片列表
    for(int i = 0; i<pics.length; i++) {
        ImageView iv = new ImageView(this);
        iv.setLayoutParams(mParams);
        iv.setImageResource(pics[i]);
         views.add(iv);
    }
    vp = (ViewPager) findViewById(R.id.viewpager);
    //初始化 Adapter
    vpAdapter = new ViewPaperAdapter(views);
    vp.setAdapter(vpAdapter);
    //绑定回调
    vp.setOnPageChangeListener(this);
    //初始化底部小点
     initDots();
   private void initDots() {
   LinearLayout ll = (LinearLayout) findViewById(R.id.ll);
     dots = new ImageView[pics.length];
     //循环取得小点图片
     for (int i = 0; i<pics.length; i++) {
       dots[i] = (ImageView) ll.getChildAt(i);
       dots[i].setEnabled(true);//都设为灰色
```

```
            dots[i].setOnClickListener(this);
            //设置位置 tag,方便取出与当前位置对应
            dots[i].setTag(i);
        }
        currentIndex = 0;
        //设置为白色,即选中状态
        dots[currentIndex].setEnabled(false);
    }
    //设置当前的引导页
    private void setCurView(int position)
    {
        if (position<0 || position >= pics.length) {
            return;
        }
        vp.setCurrentItem(position);
    }
//设置当前引导小点的选中
    private void setCurDot(int positon)
    {
        if (positon<0 || positon>pics.length - 1 || currentIndex == positon) {
            return;
        }
        dots[positon].setEnabled(false);
        dots[currentIndex].setEnabled(true);
        currentIndex = positon;

    }
    //当滑动状态改变时调用

    @Override
    public void onPageScrollStateChanged(int arg0) {
        // TODO Auto-generated method stub
    }
    //当前页面被滑动时调用
    @Override
    public void onPageScrolled(int arg0, float arg1, int arg2) {
        // TODO Auto-generated method stub
    }
    //当新的页面被选中时调用
    @Override
    public void onPageSelected(int arg0) {
        //设置底部小点选中状态
```

```
            setCurDot(arg0);
        }
    @Override
        public void onClick(View v) {
            int position = (Integer)v.getTag();
            setCurView(position);
            setCurDot(position);
    }
```

(6) SettingsActivity 设计(游戏设置界面)

背景音乐和背景音效的设置:

```
mBackMusicSwitcher = (ImageView) findViewById(R.id.back_music_switcher);
        mSoundEffectSwitcher = (ImageView) findViewById

(R.id.sound_effect_switcher);
        mSettings = getSharedPreferences(Utils.PREF_SETTINGS, 0);
        boolean backMusicOff = mSettings.getBoolean(
                Utils.PREF_SETTINGS_BACK_MUSIC_OFF, false);
        boolean soundEffectOff = mSettings.getBoolean(
                Utils.PREF_SETTINGS_SOUND_EFFECT_OFF, false);

        setBackMusicRes(backMusicOff);
        setSoundEffectRes(soundEffectOff);

        mBackMusicSwitcher.setOnClickListener(new OnClickListener() {

            @Override
            public void onClick(View v) {
                boolean flag = isBackMusicOff();
                mSettings.edit().putBoolean(Utils.PREF_SETTINGS_BACK_MUSIC_OFF,
                        !flag).commit();
                setBackMusicRes(!flag);
            }
        });

        mSoundEffectSwitcher.setOnClickListener(new OnClickListener() {

            @Override
            public void onClick(View v) {
                boolean flag = isSoundEffectOff();
                mSettings.edit().putBoolean(Utils.PREF_SETTINGS_SOUND_EFFECT_OFF,
                        !flag).commit();
```

```
            setSoundEffectRes(!flag);

        }
    });

    Utils.setAD(this);
}

private boolean isBackMusicOff() {
    return mSettings.getBoolean(Utils.PREF_SETTINGS_BACK_MUSIC_OFF, false);
}

private boolean isSoundEffectOff() {
    return mSettings.getBoolean(Utils.PREF_SETTINGS_SOUND_EFFECT_OFF, false);
}

private void setBackMusicRes(boolean off) {
    if (off) {
        mBackMusicSwitcher.setBackgroundResource(R.drawable.sound_off);
    } else {
        mBackMusicSwitcher.setBackgroundResource(R.drawable.sound_on);
    }
}

private void setSoundEffectRes(boolean off) {
    if (off) {
        mSoundEffectSwitcher.setBackgroundResource(R.drawable.sound_off);
    } else {
        mSoundEffectSwitcher.setBackgroundResource(R.drawable.sound_on);
    }
}
```

9.6 本章小结

本章通过项目案例介绍了一款闯关类游戏的设计，该项目案例使用了 Ontouch 函数和 SharedPreferences，并且，通过 Service 服务实现背景音乐的播放，以及对屏幕触摸事件的监听。

参考文献

[1] 工业和信息化部通信行业职业技能鉴定指导中心,中国移动互联网基地,等. 移动应用开发技术[M]. 北京:机械工业出版社,2012.

[2] 柳贡慧,黄先开. 移动终端应用创意与程序设计[M]. 北京:电子工业出版社,2012.

[3] 柳贡慧. 3G智能手机创意设计——首届北京市大学生计算机应用大赛获奖作品[M]. 北京:电子工业出版社,2011.